MECHANICS, HEAT, and SOUND

MECHANICS, HEAT, and SOUND

Isaac Maleh
**University of Maryland
Baltimore County**

CHARLES E. MERRILL PUBLISHING COMPANY, COLUMBUS, OHIO

A Bell & Howell Company

Merrill Physical Science Series

Robert J. Foster and Walter A. Gong, *Editors*
San Jose State College

Copyright © 1969 by CHARLES E. MERRILL PUBLISHING COMPANY
Columbus, Ohio. All rights reserved. No part of this book
may be reproduced in any form, electronic or mechanical, including photocopy, recording, or any information storage and
retrieval system without permission in writing from the publisher.

Library of Congress Catalog Card Number: 68–20655
Standard Book Number: 675–09638–3 (paperback edition)
675–09638–1 (casebound edition)

1 2 3 4 5 6 7 8 9 10 – 73 72 71 70 69
PRINTED IN THE UNITED STATES OF AMERICA

Editors' Foreword

As curricula become more crowded in this age of rapidly expanding knowledge and specialization, more and more colleges and universities are turning to integrated interdisciplinary courses to transmit the basic essentials of science to non-science majors. We believe that the rigid structure of most physical science textbooks has imposed severe limitations on instruction in these courses. Far too often, instructors trained in various specialities have had to attempt to fit the wide range of goals, abilities, and backgrounds of their students to a textbook, when the converse, of course, would be much more satisfactory.

In January, 1965, the editors, five authors, and representatives of Charles E. Merrill Publishing Co., met in San Francisco to implement a new conception of physical science textbooks. The result is the *Physical Science Series,* a collection of specially written, integrated materials in short, paperback form for the college physical science program. Our coordinated efforts were directed by three vital principles.

1. The Series permits maximum flexibility of use by instructors and students. Each paperback textbook represents a five-to-seven-week section of instruction, and may be used in any sequence or combination desired by the instructor. In addition, freedom of sequence within a single book is possible. This flexibility is especially helpful in courses that include laboratory experience. In this way it is hoped that each instructor will be free to choose the most appropriate materials for his students.

2. The subject areas are portrayed in a valid manner. Each book is written by a specialist in a different discipline—physicist, chemist, astronomer, meteorologist, geologist, and science educator. Thus, in place of a homogeneous blend of textbook statements, the individual paperback textbooks have distinctive scientific flavors. The student can discover both the contrasts and underlying unities in the viewpoints of scientists in different disciplines; he can, for example, compare the approach of the physicist, who

performs laboratory experiments, with that of the geologist, who depends largely on observations of natural occurrences.

3. Scientific communication is clear, concise, and correct. Each author is both academician and experienced teacher. He has designed instruction around carefully selected scientific principles logically related to laws, definitions, and associated phenomena. Technology is used to provide illustrative examples rather than a myriad of facts to be remembered. Mathematical reasoning is used only when the sciences are made more (not less) understandable for the non-science major. Scientific jargon and excessive nomenclature are avoided.

San Jose, California *Robert J. Foster*
November, 1965 *Walter A. Gong*

Table of Contents

Chapter 1
THE DISCRIPTION OF MOTION   1

  1-1 MOTION: BEFORE GALILEO   1
          The Aristotelian Theory of Motion, 1
          Downfall of the Aristotelian Theory of Motion, 7

  1-2 MOTION: AFTER GALILEO   10
          Distance and Velocity, 10
          Distance, Velocity, and Acceleration, 11
          Is Motion Even More Complex? 13
          Vectors, 14
          An Understanding of Motion, 17

  1-3 CIRCULAR MOTION   19

  1-4 REVIEW   21

PROBLEMS   22

Chapter 2
NEWTON'S LAWS OF MOTION   23

  2-1 INTRODUCTION   23

  2-2 FIRST LAW OF MOTION   24
          Statement and Analysis, 24
          Comments:
             Concept of Inertial Frame of Reference, 25
             Examples of Inertial and Non-Inertial Frames, 27
             Concept of System, 29

2-3	SECOND LAW OF MOTION
	Statement and Analysis, 29
	A Unit of Force, 31
	Comments:
	   Direction of Force, Direction of Velocity,
	     and Direction of Acceleration, 32
	   The Words "Net," "External," and "On," 32
		29

2-4	THIRD LAW OF MOTION Statement and Analysis, 33	33
2-5	MOTION: ARISTOTELIAN VERSUS NEWTONIAN The Importance of Initial Conditions, 35	34
2-6	REVIEW	36
PROBLEMS		37

Chapter 3
NEWTON'S LAW OF GRAVITATION   39

3-1	INTRODUCTION	39

3-2	THE GRAVITATIONAL FORCE
	Background, 39
	Statement and Analysis, 40
	G, The Universal Gravitational Constant, 43
	   Why G? 43
	   The Measurement of G, 43
		39

3-3	EXAMPLES OF THE GRAVITATIONAL FORCE	44
3-4	SOURCE OF THE GRAVITATIONAL FORCE Action-at-a-Distance or Gravitational Field, 47	47
3-5	REVIEW	49
PROBLEMS		49

Chapter 4
A COUPLING OF NEWTON'S LAWS OF MOTION AND NEWTON'S LAW OF GRAVITATION: I   50

4-1	INTRODUCTION	50
4-2	THE PRINCIPLE OF EQUIVALENCE Inertial Mass and Gravitational Mass, 50	50

Table of Contents   ix

    4-3   MOTION ON THE EARTH   51
            Falling Bodies and the Gravitational Force, 52
            Motion Due to Forces Other Than Gravitational, 52

    4-4   MOTION IN THE SOLAR SYSTEM   53
            The Solar System and Johannes Kepler, 53
            The Solar System and Isaac Newton, 55
            Importance of Initial Conditions, 57
            Consequence of the Principle of Equivalence, 58

    4-5   THE MOON AND MAN-MADE SATELLITES   58

    4-6   NEWTON'S LAWS AND KEPLER'S LAWS—A COMPARISON   59

    4-7   REVIEW   60

PROBLEMS   61

Chapter 5
A COUPLING OF NEWTON'S LAWS OF MOTION AND NEWTON'S LAW OF GRAVITATION: II   **63**

    5-1   INTRODUCTION   63

    5-2   WEIGHT AND WEIGHTLESSNESS   63
            The Sensation of "Free Fall," 63

    5-3   TIDES   68
            Without Mutual Rotation, 68
            With Mutual Rotation, 69

    5-4   PERTURBATIONS OF THE PLANETARY ORBITS   70
            The Discovery of Neptune and Pluto, 72
            Mercury and General Relativity, 72

    5-5   ODDITIES OF GRAVITATION   73
            Speed of the Gravitational Force, 73
            Saturation of the Gravitational Force, 73

PROBLEMS   75

EPILOGUE   **76**

A BRIEF HISTORY OF MOTION AND GRAVITATION   76

WHAT DO WE MEAN BY "EXPLAIN"?   78

Chapter 6
CONSERVATION LAWS — 81

- 6-1 INTRODUCTION — 81
- 6-2 AN EXAMPLE OF CONSERVATION — 81
  A Game of Poker, 81
  Important Aspects of Conservation Laws, 82
- 6-3 CONSERVATION OF MOMENTUM — 83
  Definition of Momentum, 83
  Statement and Analysis, 83
  Examples of the Conservation of Momentum, 84
  Use of Conservataion of Momentum, 87
- 6-4 CONSERVATION OF ANGULAR MOMENTUM — 89
  Definition of Angular Momentum, 89
  Statement and Analysis, 89
- 6-5 CONSERVATION OF ENERGY — 90
  Definition and Remarks on Energy, 90
  Statement and Analysis, 91
  Potential Energy, 92
  Kinetic Energy, 93
  Example of Conservation of Energy:
    A Simple Pendulum, 94
- 6-6 HEAT AS A FORM OF ENERGY — 97
- 6-7 REVIEW — 99
- PROBLEMS — 99

Chapter 7
THERMODYNAMICS — 101

- 7-1 INTRODUCTION — 101
- 7-2 THE COMPOSITION OF MATTER — 101
  Elements, Compounds, Mixtures, 103
- 7-3 THE ATOMISTIC NATURE OF MATTER — 104
  Atoms and Molecules, 104
  Proof of the Atomistic Nature of Matter:
    John Dalton of Manchester, England, 106
    Amadeo Avogadro of Turin, Italy, 108
  Avogadro's Number; Definition of a Mole:
    How much does a molecule weigh? 112

Table of Contents xi

7-4 STATES OF MATTER 112
    A Description of the States of Matter:
      The Solid State, 113
      The Gaseous State, 113
      The Liquid State, 113
    Characteristics Which Determine the States of Matter:
      Molecular Composition, 114
      Pressure, 115
      Temperature, 115

7-5 TEMPERATURE 115
    The Beginning of Thermometry:
      The Galilean Thermometer, 116
    Thermometers and Temperature Measurement:
      Development of the Mercury Thermometer, 117
      Degrees Fahrenheit, Degrees Centigrade, 118
      Absolute Zero, the Kelvin Temperature Scale, 119

7-6 HEAT AND TEMPERATURE 121

7-7 REVIEW 122

PROBLEMS 122

Chapter 8
THE KINETIC THEORY OF GASES 124

8-1 INTRODUCTION 124

8-2 EXPERIMENTAL FACTS CONCERNING GASES 124
    The Perfect Gas Law, 124
    Development of the Perfect Gas Law:
      Boyle's Law, 126
      Charles's Law, 126

8-3 THE KINETIC THEORY OF GASES 127
    Model of an Ideal Gas, 127
    Deductions from the Model, 128

8-4 AN INTERPRETATION OF THE TEMPERATURE OF A GAS 130

8-5 REVIEW 131

PROBLEMS 132

Chapter 9
SOUND 133

9-1 INTRODUCTION 133
9-2 WAVES 133
    Vocabulary and Characteristics, 133
    Tranverse Waves and Longitudinal Waves, 137
9-3 SOUND 138
    Frequency Range of Sound, 138
    Velocity of Sound in Air, 139
    Production of Sound, 139
9-4 A VIBRATING STRING 140
    Frequency of a Vibrating String, 140
    Fundamental Frequency, Harmonic Frequencies, 141
    Importance of Harmonic Frequencies, 144
9-5 THE EYE AND THE EAR 145
    Electromagnetic Waves and the Eye;
    Sound Waves and the Ear, 145
9-6 REVIEW 146
PROBLEMS 146

Appendix 1
A FALLING BODY 147

Appendix 2
VECTOR ADDITION AND SUBTRACTION 149

Appendix 3
PROBLEMS WITH THE CONCEPT OF A STRAIGHT LINE 151

INDEX 155

INTRODUCTION

As stated in the title, this book is concerned with the subjects of mechanics, heat, and sound.

The subject of mechanics is at least 2000 years old. Mechanics involves the motion of matter through space and time, sometimes with the application of forces and sometimes without the application of forces. The study of mechanics began with an attempt to understand both the motion of earthly bodies and the motion of heavenly bodies. At first, the earth was thought to be fixed and immovable in the center of the universe, so it seemed reasonable to think that motion on earth was separate and distinct from motion in the heavens. Thus, there were separate theories covering each of these domains. When the earth came to be thought of as only one of many bodies orbiting the sun, however, the earth lost its uniqueness. There was no longer any reason to think that motion on the earth was basically different from motion in the heavens. That is, one theory of motion should incorporate both motion on earth and motion in the heavens.

In order to discover this theory of motion, an adequate description of motion had to be developed. This description involves a certain vocabulary (distance, velocity, acceleration, vectors) and is considered in the first chapter of this book. Armed with an adequate description of motion, we can discuss the theory of motion as contained in Newton's laws of motion and Newton's law of gravitation. The second chapter discusses Newton's laws of motion utilizing the vocabulary developed in the first chapter as well as the concepts of force, mass, and inertial frames of reference. These laws of motion describe the behavior of bodies with and without forces being exerted on them. However, in order to understand the motion of the planets as well as the motion of falling bodies on the earth, Newton presented his law of gravitation. That is, Newton introduced a hypothetical force—the so-called Gravitational Force (Chapter 3). By coupling the laws of motion with the gravitational force, Newton was able to "explain" the motion of heavenly bodies as well as the motion of earthy bodies (Chapters 4 and 5).

The problem of mechanics seemed to be solved. However major difficulties appeared in 1905 and in 1917 with the advent of Einstein's Relativity Theories. According to these theories, time does not flow at the same rate for all bodies and all observers; also the length of an object varies with the observer. According to these theories, there is no gravitational force, but rather a four-dimensional space-time geometry through which the earth is moving. The shape of this geometry is determined by the presence of masses. Unfortunately, there is not enough space in this book to delve into all of these latter developments in a satisfactory manner. The relativity theories and the problems they introduce into Newtonian mechanics are briefly sketched in Section 2.2, the Epilogue, and in Appendix 3.

The subject of heat (or, more properly, the subject of thermodynamics), involves matter, energy, and their interrelationships (Chapters 7 and 8). It is a very broad subject, hence our discussion is limited. After a preliminary description of the structure of matter, the measurement of temperature, and the conceptual difference between heat and temperature, the so-called Kinetic Theory of Gases is discussed. This theory is an attempt to understand the actual behavior of a gas by means of a model. It is shown that in spite of the faults of our model, it does supply us with a reasonably satisfactory explanation of the behavior of a gas.

The subject of sound is covered in Chapter 9. Sound is a certain kind of wave, so characteristics of waves are considered before delving into the peculiarities of sound itself.

I would like to thank Nathan Feifer for his criticism of an early version of the manuscript and Walter Gong for his advice and criticism of the manuscript. I would also like to thank Mrs. Willow Taylor for the typing of this work.

MECHANICS, HEAT,
and SOUND

MECHANICS

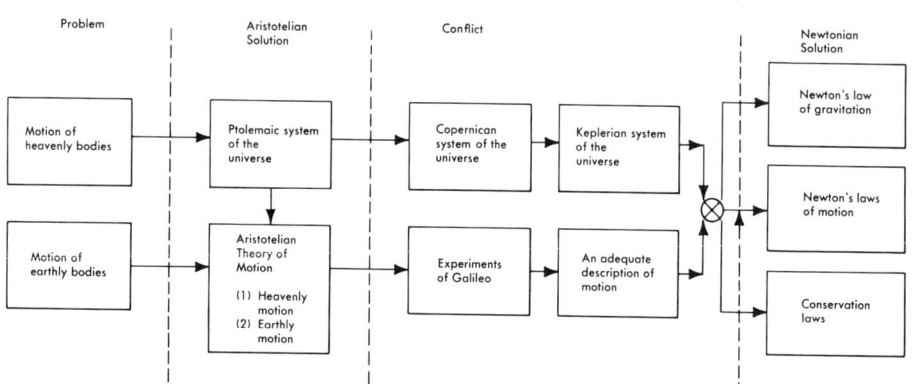

However:

Basic to the Newtonian solution of motion is the idea that space and time are independent of and separate from each other. That is, according to the Newtonian solution, matter moves in three-dimensional space while time moves uniformly onward, independent of the motion of the matter and independent of the observer. The relativity theories (circa 1905) showed that space and time are not independent of each other. So while the Newtonian theories are very useful, they are only an approximation to the truth.

Chapter 1

The Description of Motion

1-1 MOTION: BEFORE GALILEO

The Aristotelian Theory of Motion

Prior to 1550, man's view of the universe and motion within the universe could be summarized as follows.* A) The earth is fixed in the center of the universe, and all of the planets, the sun, and the stars rotate about the earth. This is called the *Ptolemaic system* of the universe. (See Fig. 1-1.) B) In keeping with assumption A, all bodies in the universe are divided into two parts, earthly bodies and heavenly bodies. Also, the physical laws (forces and laws of motion) for earthly bodies are different from the physical laws for heavenly bodies.

Earthly bodies	**Heavenly bodies**
1. All earthly bodies are composed of four basic "elements." These basic elements are earth, water, air, and fire. For example, a stone is composed mostly of earth, some water, a little air and very little fire, while a	1. The phenomena of the heavens are beautiful. They are also complicated, so only a brief description will be presented. All of the heavenly bodies; i.e., the stars, the planets, the sun, and the moon, rotate in a circle about

---

* There were many theories concerning nature, the universe, and motion that existed prior to 1550. Since this is a physics text, and not a history text, I have not attempted to present all the theories in an ordered fashion. Rather, I have taken something from each theory in order to provide a closed set of ideas and to give the student an impression of how people viewed the universe prior to Copernicus and Newton.

Mechanics, Heat, and Sound

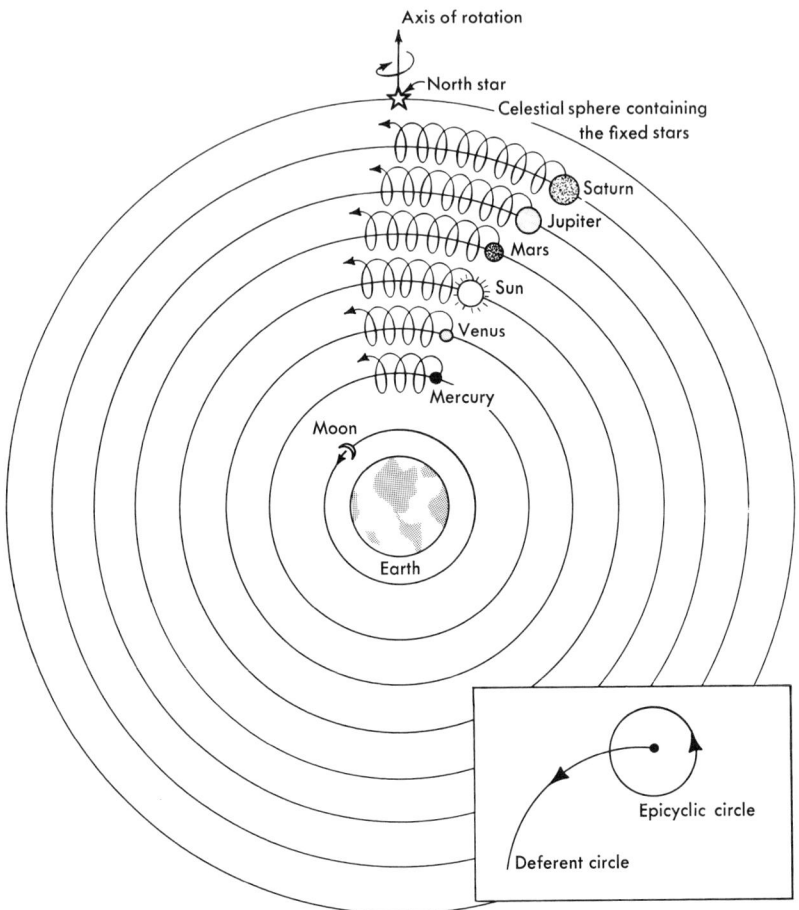

The planets move in epicycles about the earth. That is, the planets move in small circles the centers of which move in a large circle about the sun. The small circle is called an *epicycle*, the large circle is called a *deferent*. Diagram is exaggerated

Fig 1-1. Schematic diagram of the Ptolemaic system of the universe.

Earthly bodies	Heavenly bodies
gas is composed mostly of air, some fire, a little water and very little earth. (See Fig. 1-2.)	the North Star once every day.* This fact is not entirely obvious for two reasons: (1) the brilliance of the sun masks the circular motion

---

* Actually, the stars rotate almost 361° every day instead of 360°; so the stars undergo a small net rotation of about 1° each day.

The Description of Motion

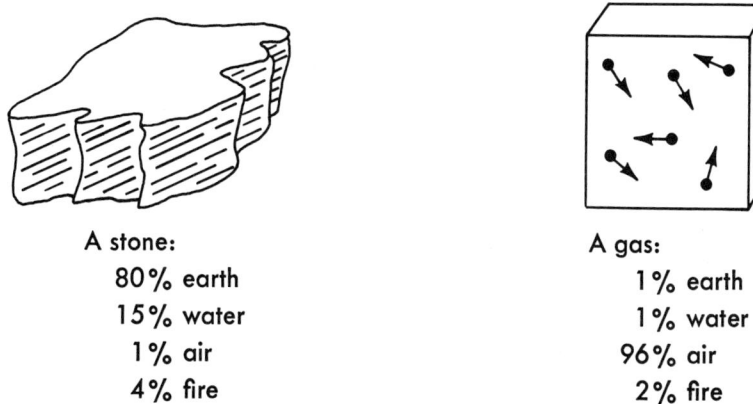

Fig. 1-2. One theory concerning the composition of matter.

Earthly bodies	Heavenly bodies
♦	of the stars and planets during the daylight hours, and (2) the earth's horizon cuts off the lower portion of the circle traced out by the sun, the moon, and various stars and planets. This cutting off results in our seeing these bodies "rising" or "setting" instead of completing their circular route about the North Star.
♦	In addition to this daily rotation of all the heavenly bodies about the North Star, the planets, the sun, and the moon move in a peculiar fashion relative to each other and relative to the stars. But in spite of all this motion, at the end of one year, the sun and the stars are in exactly the same position as they were the year before. The moon and the planets, however, do not have this yearly periodicity.
	The daily rotation of all of the heavenly bodies as well as the peculiar motion of the planets, the

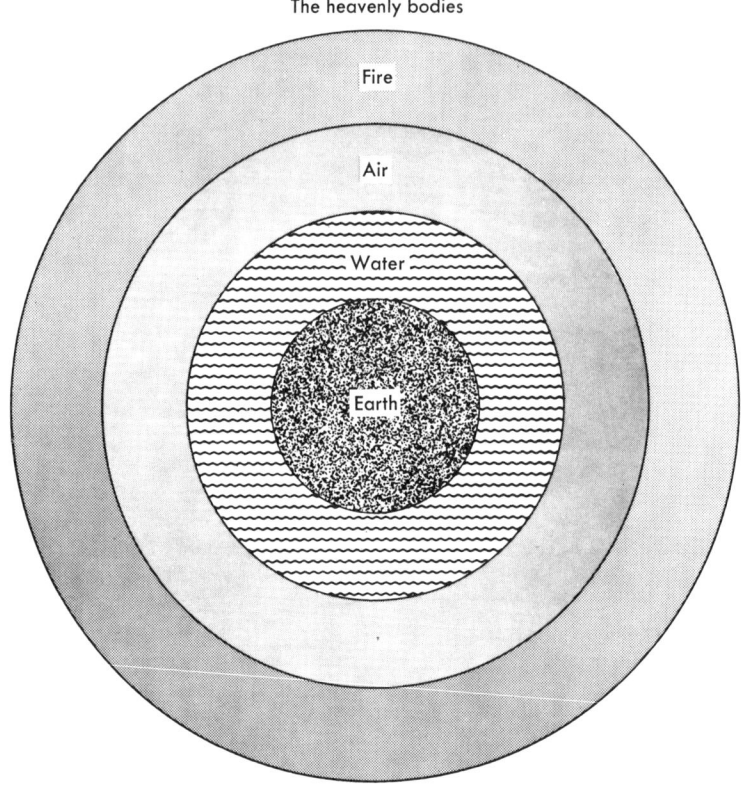

Fig. 1-3. The natural position of the elements.

Earthly bodies	Heavenly bodies
	sun, and the moon, relative to each other and relative to the stars, fascinated ancient man. He explained them using the following principles.
2. Each element has a "natural" position. For example, the natural position of the element earth is in the center of the universe, while the natural position of the element water is on top of the element earth. Air prefers to lie on top of water and fire on top of air. (See Fig. 1-3.)	2. All heavenly bodies are composed of special substances and have an innate intelligence or soul. The souls of heavenly bodies are superior to those of man, and so the natural position of these bodies is in the heavens.

The Description of Motion

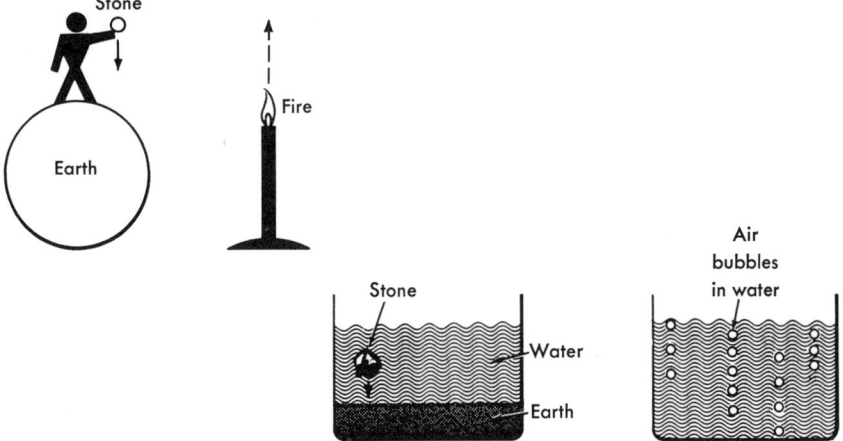

Fig. 1-4. Natural motion of earth, water, air, and fire.

Earthly bodies	**Heavenly bodies**
3. There are two kinds of earthly motion, "natural" and "unnatural."	3. The natural motion of heavenly bodies is circular because the circular form is beautiful, symmetrical, and perfect.
(a) Natural motion: if a body is in its natural position, it will remain at rest. If a body is not in its natural position, it will move toward its natural position. No external forces are necessary for a body to move toward its natural position. For example: a stone, because it is earth, would fall down toward the center of the earth. Fire rises because its natural position is above air. An air bubble in water rises because air prefers to lie on top of water. (See Fig. 1-4.)	♦
(b) Unnatural motion: All motion other than natural motion is unnatural motion, and requires a so-called *moving force*. For example: An arrow or a cannonball, composed mostly of earth, should move straight down toward the center of the earth.	♦

Earthly bodies	Heavenly bodies
However, these objects sometimes move horizontally as well as downwards and so, for the horizontal portion of the motion, a moving force is needed. In fact, any horizontal motion requires a moving force. (See Fig. 1-5.)	◆
◆	4. To explain the phenomena described in item 1 using the principles in items 2 and 3, the stars were located on a distant, celestial sphere and the sun, moon, and the planets were located on spheres in between this celestial sphere and the earth. The celestial sphere had an axis through the North Star and was thought to rotate about this axis once every day dragging the inner spheres around with it.* This daily rotation provided every star, the sun, the moon, and the planets with one basic circular orbit. However, to explain the peculiar motion of the planets, the sun, and the moon relative to each other and relative to the stars, two additional circular orbits were attributed to each of the planets (called deferent and epicyclic circles) and one additional circular orbit was attributed to the sun and the moon. (See Fig. 1-1.)
◆	According to this theory, no forces were necessary for this natural motion of the sun, moon, planets, and stars.

This scheme of the universe, besides being relatively simple, was subtly satisfying. Man's ego was simultaneously inflated and deflated. Inflated, because he was in the center of the universe and the whole world rotated around him. Deflated, because the motion associated with man and his environment

---

* Actually, the celestial sphere was thought to rotate almost 361° every day. This was to account for the small net rotation that the stars undergo every day. See the footnote on page 2.

The Description of Motion

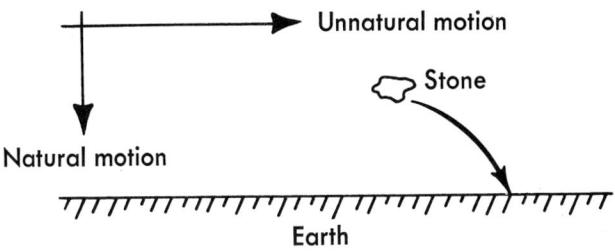

Fig. 1-5. **Combination of natural and unnatural motion.**

was neither beautiful nor perfect, while the motion of heavenly bodies was both beautiful and perfect. Since man's personality is filled with subtle contradictions, this scheme of the universe is in harmony with man's personality.

But there were difficulties.

Although the theory of heavenly motion could not be tested directly, as long as it agreed with the observations, there was no reason to doubt it. However, the theory of earthly motion was slightly inadequate.

An arrow is composed mostly of earth, so its natural position is in the center of the earth and its natural motion is down, toward the center of the earth. But an arrow, once it has left its bow, does not exhibit natural motion. It moves in some sort of arc which is a combination of natural and unnatural motion. Unnatural motion, however, requires a moving force and there is nothing touching the arrow, so where does the moving force come from?

It was thought by many people that as the arrow (or cannonball, or any moving object) moved, it created a vacuum just behind itself. "Nature abhors a vacuum," so air moved in to fill the vacuum, and in so doing pushed the body further along. It was just a problem of mathematics to understand unnatural motion. One had to compute exactly how the air rushed in to fill the vacuum. It depended on the shape of the object, its speed, etc. The question of motion within a vacuum did not even arise, because there was no such thing as a vacuum.

In spite of the weakness of these arguments, some people felt that earthly motion was understood, at least qualitatively if not quantitatively.

Downfall of the Aristotelian Theory of Motion

Two major discoveries promoted the downfall of the Aristotelian theory of motion. One discovery was the realization that the sun, not the earth, is at the center of the universe, and that the earth is only one of many planets orbiting the sun. The second discovery was the realization of an almost unbelievable fact: a 10-lb rock and a 1-oz ball, when dropped from the same height, will reach the ground at the same time.

Let's consider each discovery in turn.

1. In 1543, the Copernican system of the universe was presented. About 60 years later, this system evolved into the so-called Keplerian system of the universe. According to the Keplerian system, the sun is in the center of

the universe and all the planets, including the earth, are revolving around the sun in elliptical orbits. The earth, in addition to its revolving around the sun, is also rotating about its north-south axis once each day. Finally, the fixed stars lie on a sphere surrounding the furthest planetary orbit. (See Fig. 1-6.)

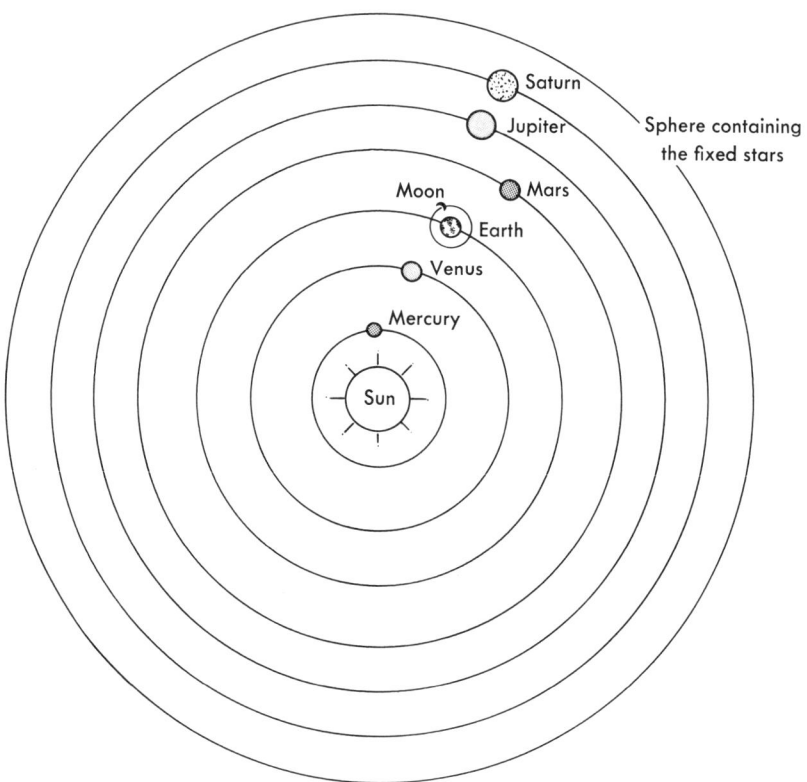

Fig. 1-6. Schematic diagram of the Copernican and Keplerian systems of the universe.

Using the Keplerian system rather then the Ptolemaic system, the phenomena of the heavens are easier to understand. Why do the heavenly bodies appear to rotate daily about the North Star? The earth is rotating about its north-south axis once every day and this axis passes through the North Star; it is the earth that is rotating, not the heavenly bodies. Why is it that, at the end of one year, the sun and the stars are in exactly the same position as they were the year before? The earth is revolving around the sun with a period of one year. The Keplerian system, without resorting to a complex system of epicyclic and deferent circles, also explained the peculiar motion

The Description of Motion

of the planets, the sun, and the moon relative to each other and relative to the stars.

Unfortunately, although the phenomena of the heavens were easier to understand, the problem of motion reappeared. Why do the planets orbit the sun in ellipses rather than in circles? The earth is only one of several planets and there is no reason to believe that the substances comprising the other planets are different from those comprising the earth; i.e., the planets contain earthly matter. But if the planets contain earthly matter, the planets should fall toward the earth. They do not! If the sun is in the center of the universe, why don't the planets fall into the sun? Etc., etc.

Hereafter, we will call the sun and all the planets orbiting it, the *solar system*.

2. If one simply takes a heavy object and a light object* and drops them both from a height of say, ten feet, both will reach the ground at the same time. According to the older theory of motion, since the heavier object contains more "earth" than the lighter object, the heavier object should reach the ground first. In fact, if it is twice as heavy, it should reach the ground in half the time. But they both reach the ground at the same time.† This "ball-dropping" type of experiment is attributed to Galileo in the early 1600's. Galileo (1564-1642) conducted many experiments on motion and is chiefly responsible for the analysis of motion presented in the following section. (See Fig. 1-7.)

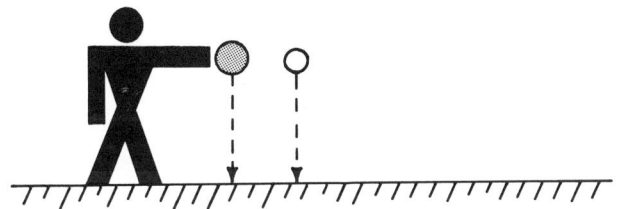

The heavy ball and the light ball will reach the ground at the same time.

Fig. 1-7. "Ball-dropping" experiments.

---

* Because of the existence of air, two things must be taken into consideration; namely, air friction and the buoyancy of air. The shape of the objects should be the same so that the air friction is the same for both objects, and the light object should not be too light, because the air will buoy it up; i.e., don't use a feather. Also, winds might affect the result. However, if air were absent, a 10-ton truck and a feather would reach the ground at the same time.

† This startling fact has produced many interesting results. In the early 1600's it dealt a crushing blow to the Aristotelian theory of motion. But by the end of the 1600's this fact was understood in the light of Newton's laws of motion and Newton's law of gravitation. According to this explanation, an esoteric concept called the *principle of equivalence* was responsible for this fact. Three hundred years later, in the early 1900's, Einstein used this concept to help develop his general theory of relativity. For a more complete discussion, see *Modern Physics,* I. Maleh, Charles E. Merrill Publishing Co., Columbus, Ohio, 1966.

In view of these facts concerning heavenly motion and earthly motion; i.e., the sun-centered universe and the result of the ball-dropping experiments, the existing theories of motion had to be modified. It soon became apparent that any theory of motion must be equally applicable to both heavenly motion and earthly motion. That is to say, any theory of motion must be able to explain the motion of all of the planets in the solar system as well as the motion of all the bodies on the earth.

1-2 MOTION: AFTER GALILEO

Although we really want an adequate theory of motion, we must first have an adequate description of motion. It took the combined efforts of many men to formulate an adequate description of motion, and afterward, an adequate theory of motion. These formulations were predominantly those of Galileo, Descartes, Huygens, and Newton. In order to understand the resulting theory of motion, applicable to all bodies and not confined to either earthly bodies or heavenly bodies, let us go through some simple experiments to find out what words and concepts are necessary to describe motion.

Distance and Velocity

As shown in Fig. 1-8, if I roll a marble on a table top, the motion of the marble can be described by the words *distance, time,* and *velocity*. First, I define velocity as the change in distance, or the distance traversed, divided by the time. If the velocity is constant, the distance traversed by the marble equals its velocity multiplied by the time required for the marble to traverse that distance. This observation can be expressed as

$$d = vt \qquad (1\text{-}1)$$

where

        d is the distance traversed
        v is the velocity of the marble
        t is the time of travel.

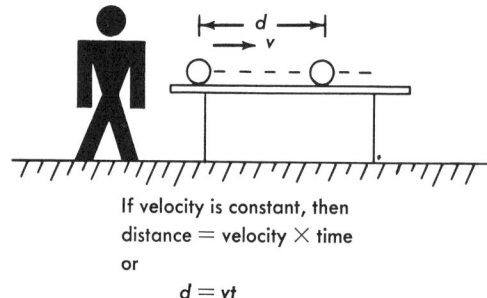

Fig. 1-8. **Vocabulary of motion—motion with constant velocity.**

The Description of Motion

Example 1-1:

If the velocity of a car is 30 m/sec, how far does it travel in 6 sec? (Assume that the velocity is constant.)

$v = 30 \dfrac{m}{sec}$

$d = vt$
$= 30 \times 6$
$= 180 \, m$

Distance, Velocity, and Acceleration

The description of motion immediately becomes more complicated if, as shown in Fig. 1-9, I drop the marble from a height, h. Now it is clear that the velocity of the marble changes as it falls towards the earth. Immediately after leaving my hand, it has a small velocity; in fact, zero velocity. As the marble approaches the ground, its speed increases and it finally strikes the ground with a large velocity.

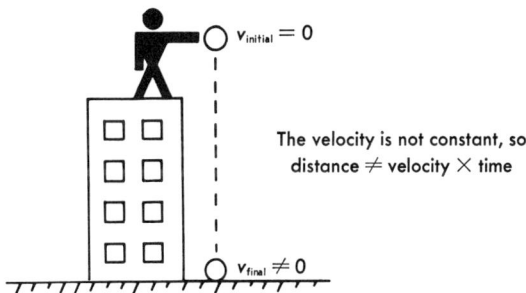

Fig. 1-9. Vocabulary of motion—the need for the concept of acceleration.

We are interested in finding an expression for the distance traversed by the marble. Since it is not clear whether to use the final velocity, the initial velocity, or some intermediate velocity, it is no longer possible to apply Eq. (1-1) directly.

We need something to describe a change in velocity. To fill that need, the concept of *acceleration* was introduced. As I previously defined velocity in terms of distance and time, let me now define acceleration in terms of velocity and time. Previously, I defined velocity as the change in distance divided by the time. Now I define acceleration as the change in velocity divided by the time.

Since the change in a quantity is equal to the final quantity minus the initial quantity, the change in velocity equals the final velocity minus the initial velocity. Expressing this information in a formula, we have:

$$a = \frac{v_{\text{final}} - v_{\text{initial}}}{t} \quad (1\text{-}2)$$

where

a is the acceleration
v_{final} is the final velocity
v_{initial} is the initial velocity
t is the time involved.

Example 1-2:
A car increases its velocity from 10 m/sec to 25 m/sec in 5 sec. What is the acceleration of the car?

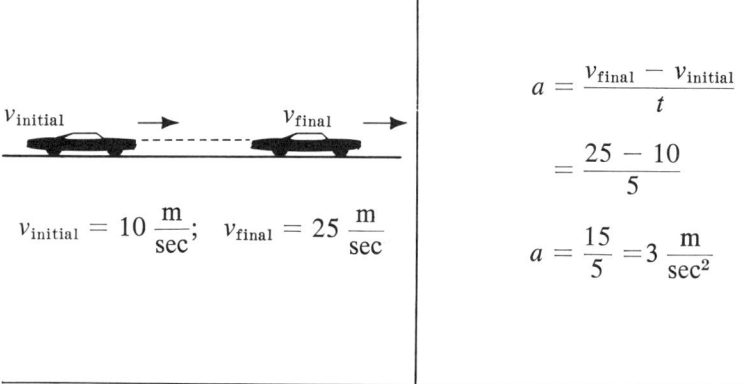

As shown in Appendix 1, by using the idea of an average velocity and the concept of *acceleration,* we can formulate an expression for the distance traversed by the marble in terms of its initial velocity, its acceleration, and the time.

The final expression for the purely vertical motion of the marble is

$$h = v_{\text{initial}} t + \tfrac{1}{2} a t^2 \quad (1\text{-}3)$$

where

h is the distance traversed
v_{initial} is the initial velocity
a is the acceleration
t is the time of travel.

The Description of Motion

Example 1-3:
If a marble is dropped with an initial velocity of 0 m/sec and it has an acceleration of 10 m/sec² downwards, how far does it go in 3 sec?

$$v_{initial} = 0 \ \frac{m}{sec}$$
$$a = 10 \ \frac{m}{sec^2}$$
$$t = 3 \ sec$$

$$h = v_{initial} t + \tfrac{1}{2} a t^2$$
$$= 0 \times 3 + \tfrac{1}{2} \times 10 \times 3^2$$
$$= 0 + \tfrac{1}{2} \times 10 \times 9$$
$$h = 45 \text{ m}$$

We have developed this equation for a marble. Is it also valid for a rock, a book, a truck, or a human being? Yes! It is amazing but true that the acceleration of a rock, a book, a truck, or a human being, falling toward the earth, is the same (10.0 m/sec²), so that Eq. (1-3) is valid for any falling object.* In fact, one major reason why the concept of acceleration is so useful is because the acceleration of all falling bodies is the same.

Is Motion Even More Complex?

Motion seems to be getting more and more complex. For purely horizontal motion, we obtained a very simple expression, $d = vt$, for the distance traversed by a body. For purely vertical motion, we obtained a more complicated expression, $d = v_{initial} t + \tfrac{1}{2} a t^2$, for the distance traversed by a body. Suppose, now, that I really complicate things.

Suppose that instead of simply dropping the marble, I throw it horizontally as shown in Fig. 1-10. From our previous experiments, we would think that an expression for the distance traveled in a certain time can be obtained by considering the time involved, the acceleration, and the average velocity. The distance traveled would be the length of the arc d_1, the time involved would be t_1, and the acceleration would be denoted by a_1.

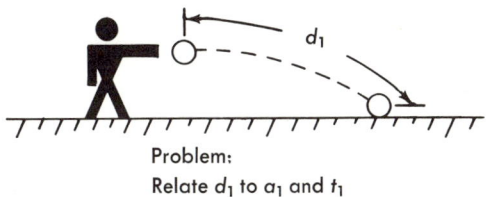

Problem:
Relate d_1 to a_1 and t_1

Fig. 1-10. **A marble is thrown horizontally.**

---

* We are not considering objects more than 100 miles above the surface of the earth; i.e., we are excluding astronauts, satellites, etc. Also, since the real number, 9.8, is so close to 10.0, and since the number 9.8 is so much more cumbersome to use than 10.0, we will generally use the number 10.0 m/sec² for the acceleration due to gravity.

In theory, the analysis is exactly the same; the situation is just more complicated because the marble is moving in an arc. We start by looking for an average velocity, introduce an acceleration, etc., etc. However, we must be prepared for complex mathematics; so before we begin, we relax and try to think of a way of avoiding the problem.

Perhaps there is a way in which we can check this idea, just to see if we are on the right track. Yes, fortunately, we can check this idea by simultaneously dropping a marble and throwing a marble horizontally. From the remarks made above, the two marbles should reach the ground at different times, because the path lengths are different. In Fig. 1-11, marble A

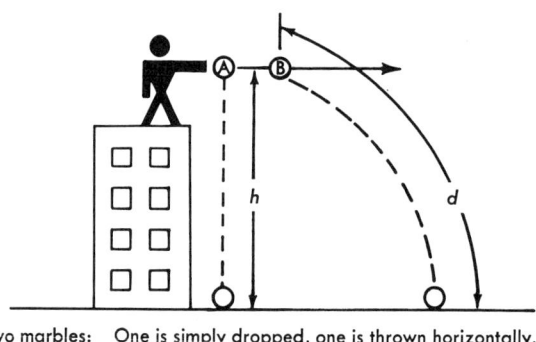

Two marbles: One is simply dropped, one is thrown horizontally.

Result: Both marbles reach the ground simultaneously.

Fig. 1-11. **"A behavior hardly comprehensible to me." Galileo, *circa* 1638.**

traverses a distance h, while marble B traverses a distance d. If this simple experiment is performed, a very startling result will appear.

Both marbles will reach the ground simultaneously.

You can repeat this experiment a million times; each and every time both marbles will reach the ground simultaneously.

This is very surprising, but it is true, nevertheless.

Why do both marbles reach the ground simultaneously when they obviously traverse different distances? To answer this question, we have to re-examine our ideas concerning motion.

We will see that the answer lies in the fact that distance, velocity, and acceleration are *directed* quantities; i.e., they are *vectors*. Vectors are not like ordinary numbers, and so they cannot be treated as ordinary numbers.

Vectors

In order to describe a human being you need to know his sex, height, weight, hair color (if he has any hair), eye color, etc., etc. An enormous amount of data is necessary to describe a human being. Fortunately, physics

The Description of Motion

Example 1-3:
If a marble is dropped with an initial velocity of 0 m/sec and it has an acceleration of 10 m/sec² downwards, how far does it go in 3 sec?

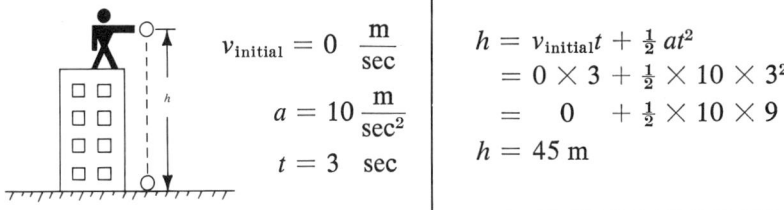

$v_{initial} = 0 \; \frac{m}{sec}$

$a = 10 \; \frac{m}{sec^2}$

$t = 3 \; sec$

$h = v_{initial}t + \frac{1}{2}at^2$
$= 0 \times 3 + \frac{1}{2} \times 10 \times 3^2$
$= 0 \quad + \frac{1}{2} \times 10 \times 9$
$h = 45 \; m$

We have developed this equation for a marble. Is it also valid for a rock, a book, a truck, or a human being? Yes! It is amazing but true that the acceleration of a rock, a book, a truck, or a human being, falling toward the earth, is the same (10.0 m/sec²), so that Eq. (1-3) is valid for any falling object.* In fact, one major reason why the concept of acceleration is so useful is because the acceleration of all falling bodies is the same.

Is Motion Even More Complex?

Motion seems to be getting more and more complex. For purely horizontal motion, we obtained a very simple expression, $d = vt$, for the distance traversed by a body. For purely vertical motion, we obtained a more complicated expression, $d = v_{initial}t + \frac{1}{2}at^2$, for the distance traversed by a body. Suppose, now, that I really complicate things.

Suppose that instead of simply dropping the marble, I throw it horizontally as shown in Fig. 1-10. From our previous experiments, we would think that an expression for the distance traveled in a certain time can be obtained by considering the time involved, the acceleration, and the average velocity. The distance traveled would be the length of the arc d_1, the time involved would be t_1, and the acceleration would be denoted by a_1.

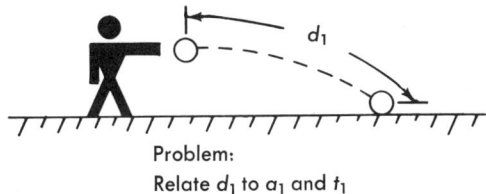

Problem:
Relate d_1 to a_1 and t_1

Fig. 1-10. A marble is thrown horizontally.

---

* We are not considering objects more than 100 miles above the surface of the earth; i.e., we are excluding astronauts, satellites, etc. Also, since the real number, 9.8, is so close to 10.0, and since the number 9.8 is so much more cumbersome to use than 10.0, we will generally use the number 10.0 m/sec² for the acceleration due to gravity.

14   *Mechanics, Heat, and Sound*

In theory, the analysis is exactly the same; the situation is just more complicated because the marble is moving in an arc. We start by looking for an average velocity, introduce an acceleration, etc., etc. However, we must be prepared for complex mathematics; so before we begin, we relax and try to think of a way of avoiding the problem.

Perhaps there is a way in which we can check this idea, just to see if we are on the right track. Yes, fortunately, we can check this idea by simultaneously dropping a marble and throwing a marble horizontally. From the remarks made above, the two marbles should reach the ground at different times, because the path lengths are different. In Fig. 1-11, marble A

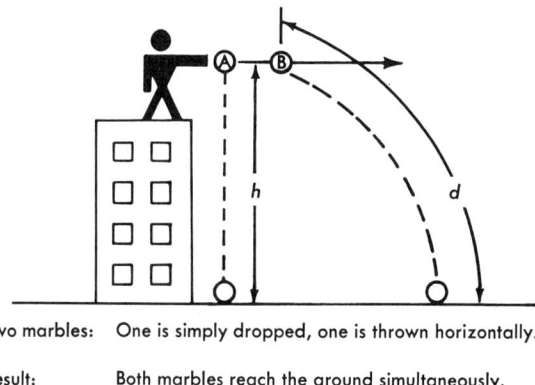

Two marbles: One is simply dropped, one is thrown horizontally.

Result: Both marbles reach the ground simultaneously.

Fig. 1-11. "A behavior hardly comprehensible to me." Galileo, *circa* 1638.

traverses a distance h, while marble B traverses a distance d. If this simple experiment is performed, a very startling result will appear.

Both marbles will reach the ground simultaneously.

You can repeat this experiment a million times; each and every time both marbles will reach the ground simultaneously.

This is very surprising, but it is true, nevertheless.

Why do both marbles reach the ground simultaneously when they obviously traverse different distances? To answer this question, we have to re-examine our ideas concerning motion.

We will see that the answer lies in the fact that distance, velocity, and acceleration are *directed* quantities; i.e., they are *vectors*. Vectors are not like ordinary numbers, and so they cannot be treated as ordinary numbers.

Vectors

In order to describe a human being you need to know his sex, height, weight, hair color (if he has any hair), eye color, etc., etc. An enormous amount of data is necessary to describe a human being. Fortunately, physics

The Description of Motion

does not deal with such complicated things, and so only a limited amount of data is necessary to describe the things of importance to physics.

We will concern ourselves with just two classes of things; one class, called *scalars*, requires only a number for its description, while the second class, called *vectors*, requires both a number and a direction for its description.

When I say that the temperature of the room is 70°F, there is nothing more to say about the temperature of the room. The listener knows all about the room as regards its temperature. When I say the volume of a refrigerator is 2 ft³, the listener knows all about the refrigerator as regards its volume. However, when I say that I walked 3 m, the listener thinks "Well, he walked 3 m, but in which direction?" Also, when I say that I'm running at 2 m/sec, the listener thinks "Yes, he is moving at the rate of 2 m/sec but in which direction?" So, while the temperature and volume require only a number for an adequate description (70°F or 2 ft³), distance and velocity require a direction as well as a number. Temperature and volume are *scalars* while distance and velocity are *vectors*. In the jargon of physics, scalars can be described by a magnitude; i.e., a number alone, but vectors require a magnitude as well as a direction for their description. (Compare this with the description of a human being.) To describe a vector, we use the following notation:

5 ∡ 45°

The physical quantity has a magnitude of 5 units and a direction of 45° relative to the horizontal.* (See Fig. 1-12 for a sampling of other vectors.)

Some examples of scalars and vectors are listed in Table 1-1.

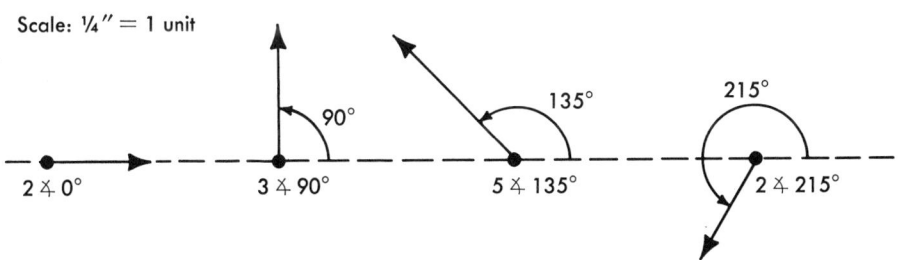

Fig. 1-12. Examples of vector notation.

While the addition and subtraction of scalars (or numbers), is relatively simple, the addition and subtraction of vectors (or directed numbers), is not

---

*In this book, we will consider only the horizontal and vertical directions; we will neglect the "sidewise" direction.

Table 1-1—Examples of Scalars and Vectors

Scalars	Vectors
Temperature	Distance
Volume	Velocity
Energy	Acceleration
Mass	Force
Time	

so simple. There are many ways of adding or subtracting vectors but perhaps the simplest way is the so-called arithmetic approach.* The arithmetic approach is to first break up each vector into a horizontal portion and a vertical portion. When all of the vectors that we are dealing with have been divided into horizontal and vertical portions, we can add or subtract the horizontal portions of each vector and we can add or subtract the vertical portions of each vector. The crucial point is that we cannot mix the horizontal portion of any vector with either the vertical portion of the same vector or with the vertical portion of any other vector.

An equation involving physical quantities which are vectors requires special interpretation. For example, the equations that describe motion require special interpretation because motion involves distance, velocity, and acceleration, and these physical quantities are all vectors. The interpretation of such an equation is actually quite simple. Since horizontal portions of vectors are only related to horizontal portions of other vectors, and vertical portions of vectors are only related to vertical portions of other vectors, only the horizontal portions of distance, velocity, and acceleration can be related

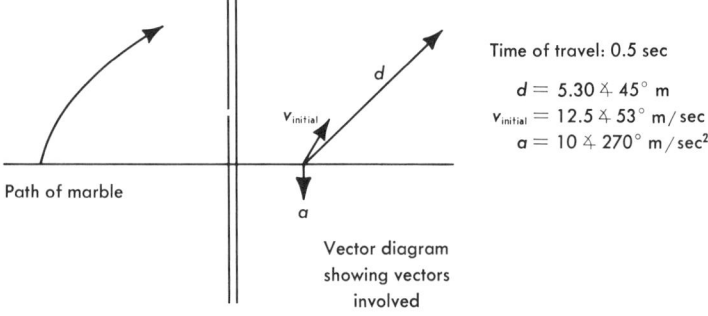

Fig. 1-13. Motion of a marble.

---

* See Appendix 2.

The Description of Motion

to each other, and only the vertical portions of distance, velocity, and acceleration can be related to each other. In particular, the horizontal portion of distance traveled involves the horizontal portion of velocity and the horizontal portion of acceleration. Similarly, the vertical portion of distance traveled involves the vertical portion of velocity and the vertical portion of acceleration. In a more succinct manner: *Vertical and horizontal motion are independent of each other.*

> For example, if the distance traversed is 5.30 ∡ 45° m (see Fig. 1-13), this means that the distance traversed is 3.75 m in the *x*-direction and 3.75 m in the *y-direction*. Also, if the initial velocity is 12.5 ∡ 53° m/sec, this means that the initial velocity was 7.5 m/sec in the *x*-direction and 10.0 m/sec in the *y*-direction. Finally, if the acceleration is 10 ∡ 270° m/sec², this means that the acceleration is 0.0 m/sec² in the *x*-direction, and − 10 m/sec² in the *y*-direction. Summarizing this data in a neater, more intelligible fashion:
>
> | **Horizontal direction** | | **Vertical direction** | |
> |---|---|---|---|
> | Distance: | 3.75 m | Distance: | 3.75 m |
> | Velocity: | $0.0 \frac{m}{sec}$ | Velocity: | $10.0 \frac{m}{sec}$ |
> | Acceleration: | $7.5 \frac{m}{sec^2}$ | Acceleration: | $10.0 \frac{m}{sec^2}$ |
>
> Now, if there is any relationship between these numbers representing horizontal and vertical portions of vectors at all, the relationship can only be among the horizontal numbers or among the vertical numbers separately. There is no relationship of the horizontal numbers with the vertical numbers. (Using $t = 0.5$ sec, see if the formula $d = vt + \frac{1}{2}at^2$ is valid for the horizontal numbers alone and for the vertical numbers alone.)

An Understanding of Motion

We are now in a position to understand why marbles A and B reached the ground simultaneously in spite of the fact that marble A was simply dropped while marble B was thrown horizontally. The key point is that distance, velocity, and acceleration are vector quantities. (See Fig. 1-14.)

While marble A is moving vertically only, marble B is moving horizontally as well as vertically. But, since vertical and horizontal motion are independent of each other, and since both marbles have the same vertical history*, both marbles must reach the ground simultaneously.

---

* By the "same vertical history" I mean they both begin from the same height; namely *h*, and they both begin with the same initial velocity in the vertical direction; namely zero.

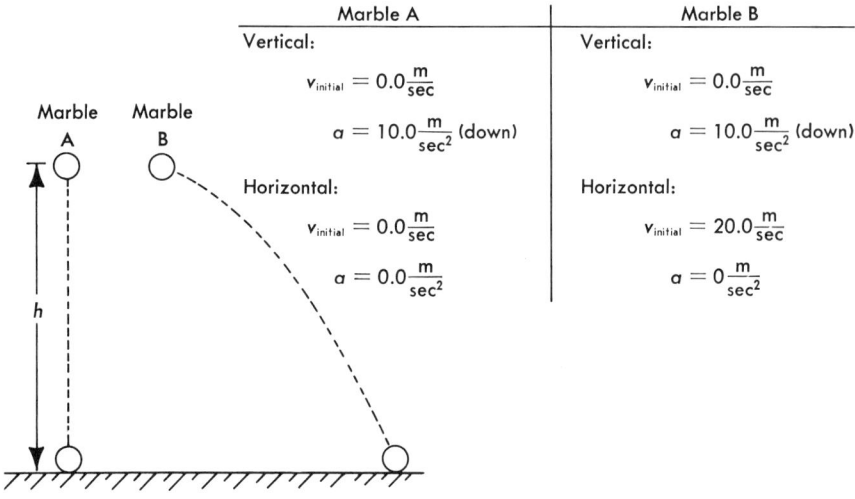

Fig. 1-14. Analyzing the motion of two marbles using vectors.

To describe the vertical descent of either marble A or marble B, we can use Eq. (1-3). This equation is applicable to the purely vertical motion of marble A as well as to the vertical portion of the motion of marble B.

Although the vertical history of both marbles is identical, the horizontal history of both marbles differ. Marble A has zero horizontal velocity, while marble B has an initial horizontal velocity of, perhaps, 20 m/sec. Since neither marble has any horizontal acceleration, each retains its initial horizontal velocity. Thus marble A retains its horizontal velocity of 0 m/sec, and marble B retains its horizontal velocity of 20 m/sec. The equation describing their horizontal motion is Eq. (1-1).

Notice, however, that Eq. (1-3) could be used for both vertical and horizontal motion if interpreted correctly.

$$\underbrace{d}_{\text{Horizontal distance}} = \underbrace{v_{\text{initial}} t}_{\text{Initial horizontal velocity}} + \underbrace{\tfrac{1}{2} \times at^2}_{\text{Horizontal acceleration}} \quad \textbf{(1-3)}$$

or

$$\underbrace{d}_{\text{Vertical distance}} = \underbrace{v_{\text{initial}} t}_{\text{Initial vertical velocity}} + \underbrace{\tfrac{1}{2} \times at^2}_{\text{Vertical acceleration}} \quad \textbf{(1-3)}$$

where the time, t, is the same for both equations. That is to say, time is the same for both the horizontal portion of the motion and the vertical portion of the motion. Time is not a directed quantity, time is a *scalar* quantity.

The Description of Motion

Example 1-4:

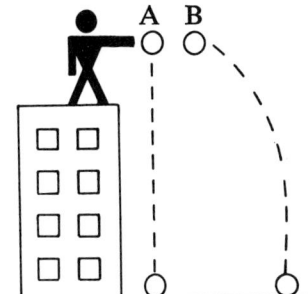

Both marble A and marble B are dropped from a building of height 180 m. However, marble B is given an initial horizontal velocity of 20 m/sec. How much time does it take marble A to reach the ground? Marble B? How far from the building will marble B fall? (Let the vertical acceleration be 10 m/sec² down.)

Marble A

Vertical motion	Horizontal motion
$d = v_{initial}t + \frac{1}{2}at^2$	$d = v_{initial}t + \frac{1}{2}at^2$
$180 = 0 + \frac{1}{2} \times 10 \times t^2$	$= 0 + 0$
$180 = 5t^2$	$= 0$
$t = 6$ sec	

Marble B

Vertical* motion	Horizontal motion
$d = v_{initial}t + \frac{1}{2}at^2$	$d = v_{initial}t + \frac{1}{2}at^2$
$180 = 0 + \frac{1}{2} \times 10 \times t^2$	$= 20 \times t + \frac{1}{2} \times 0 \times t^2$
$180 = 5t^2$	$= 20 \times 6 + 0$
$t = 6$ sec	$= 120$ m

1-3 CIRCULAR MOTION

The instructor may wish to postpone this section until after studying Chapter 3 (Newton's Law of Gravitation).

If a ball were made to go in a circle, round and round and round, offhand you would say that the ball is not accelerating. After all, its speed† is constant. If, however, you re-read the section on vectors, doubt and con-

---

* Since the vertical motions of marbles A and B are the same, it takes 6 seconds for marble B to reach the ground also.

† Speed is the word used to describe the magnitude of the velocity. Thus, if the velocity is 12 m/sec north, the speed is simply 12 m/sec.

fusion would arise. (See Fig. 1-15.) True, the speed of the ball does not change, but *speed* just represents the magnitude of the velocity. *Velocity* is a vector quantity, and it has *direction* as well as magnitude. Therefore, while

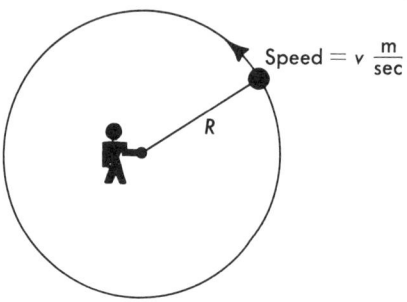

Fig. 1-15. A ball moving in a circle.

the magnitude of the velocity does not change, the direction of the velocity does. Since acceleration is defined as the change in velocity divided by the time, and a change in either the magnitude or the direction of the velocity (or both) is a change in the velocity, the rotating ball is accelerating.

What is the *magnitude* of the acceleration?
What is the *direction* of the acceleration?

This problem was originally tackled and solved by a Dutch physicist, Huygens, in 1659. He considered the problem of a man swinging a stone at the end of a sling in a circle around his head. It turns out that the magnitude of the acceleration is v^2/R, where v is the speed of the stone and R is the radius of the circular orbit. Also, the direction of the acceleration is toward the center of the circle. (See Fig. 1-16.)*

Notice the similarity of the motion of this stone to the motion of the earth

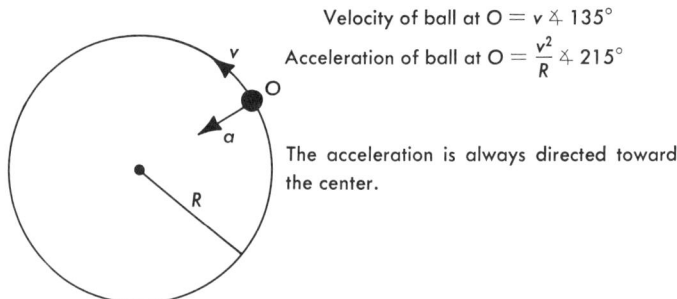

Fig. 1-16. The relationship between velocity and acceleration of a rotating ball.

---

*The proof of this statement requires the definition of acceleration, vector subtraction, the use of similar triangles, and a good deal of mathematical juggling. Fortunately, the proof is beyond the scope of this book.

The Description of Motion

orbiting the sun. Indeed, the motion of the stone is similar to the motion of any of the planets which are orbiting the sun. It is also similar to the motion of the moon and all of the man-made satellites which are orbiting the earth. (See Fig. 1-17.)

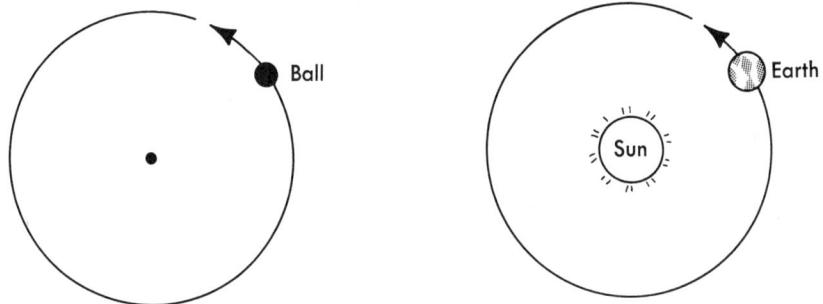

Fig. 1-17. The similarity between a rotating ball and a rotating earth.

The acceleration involved in circular motion is peculiar in that it does not affect the magnitude of the velocity at all. What exactly is the peculiarity of this sort of acceleration? It turns out that *if* the direction of the acceleration is perpendicular to the direction of the velocity, the acceleration will produce a change in the direction of the velocity only, and will not affect the magnitude of the velocity at all. In circular motion, the direction of the acceleration is toward the center of the circle and is always perpendicular to the direction of the velocity. Because the acceleration is directed toward the center of the circle, it is often called *centripetal* acceleration. The word *centripetal* means "center-seeking."

1-4 REVIEW

Although we really wanted an adequate theory of motion, we first had to have an adequate description of motion.

The concept of a vector is an important part of our description of motion. A vector is a physical quantity that requires a magnitude as well as a direction for its description. And a change in either the magnitude of a vector, the direction of a vector, or a change in both the magnitude and the direction of a vector, constitutes a change in the vector. Distance, velocity, and acceleration are all vector quantities. Velocity and acceleration are defined as follows:

$$\text{velocity} = \frac{\text{change in distance (or distance traversed)}}{\text{change in time}}$$

or

$$v = \frac{d_{\text{final}} - d_{\text{initial}}}{t} = \frac{d}{t}$$

and

$$\text{acceleration} = \frac{\text{change in velocity}}{\text{change in time}}$$

or

$$a = \frac{v_{\text{final}} - v_{\text{initial}}}{t}$$

In any problem involving motion we must realize that vector quantities are involved and, therefore, that vertical and horizontal directions are independent of each other.

Also, we must keep in mind the fact that although distance, velocity, and acceleration are all vector quantities, their similarities end there. In the same way that distance and velocity are different entities, velocity and acceleration are different entities. The difference between velocity and acceleration is most clearly illustrated in circular motion. If an object is moving in a circle, its velocity is tangential to the circle, while its acceleration is directed toward the center of the circle.

Now that we have an adequate description of motion, we need an adequate theory of motion. This theory must be able to explain the motion of all the planets in the solar system as well as the motion of all the bodies on the earth.

PROBLEMS

1. A train is traveling at a constant velocity of 8 m/sec. How far does it go in 3 minutes? [*Hint:* The answer is not 24 m.]
2. A car goes from a standing start to 60 mph in 10 sec. What is its acceleration? [*Note:* 60 mph = 27 m/sec.]
3. To determine the depth of a well, a man drops a 2-lb rock into it. (The rock has no initial velocity.) Three seconds later he sees the rock hit the bottom. How deep is the well? If he dropped a 20-lb rock, would the depth of the well appear to be deeper, shallower, or the same?
4. What is a scalar quantity? Give an example.
5. What is a vector quantity? Give an example.
6. Why are vectors necessary?
7. Two baseballs (A and B) are dropped simultaneously from two separate buildings. Each building is 180 m high. Willy Mays, star baseball player, lives half way down one of the buildings and sees ball B falling. He grabs a bat and swings, giving ball B a horizontal velocity of 28 m/sec. How long does it take ball B to reach the ground? How long does it take ball A to reach the ground? How far from the building will ball B land? [*Hints:* Draw a diagram. Equation (1-3) and the concept of vectors are the only things necessary to solve this problem. Partial answer: Ball B takes 6 sec to reach the ground and lands 49 m from the building. Use 10 m/sec² for the vertical acceleration.]
8. The earth is about 93 million miles from the sun and is moving in an approximately circular orbit with a speed of 66,600 mph. Assuming that the orbit is purely circular and that its radius is exactly 93 million miles, what is the magnitude and direction of the earth's acceleration? [*Note:* 93×10^6 mi = 150×10^9 m, and 66,600 mph = 29,800 m/sec.]
9. A man is swinging a marble about his head. If the speed of the marble is 2 m/sec and the radius of its circular orbit is 0.5 m, what is the magnitude and direction of its acceleration?

Chapter 2

Newton's Laws of Motion

In Chapter 1 we developed an adequate description of motion. In this chapter we will develop an adequate theory of motion.

2-1 INTRODUCTION

The genius of Sir Isaac Newton came forth in 1665-1666, when he combined the phenomena of motion on earth and motion in the solar system into one closed scheme. His scheme involved the concepts of velocity, acceleration, force, and inertia (or mass). Although the major portion of the work was finished in 1666, when Newton was just 23 years old, it wasn't until 1687 that he published his *Mathematical Principles of Natural Philosophy* or simply the *Principia*. It is not quite clear why Newton delayed publication. Perhaps the main reason was that he was sensitive to criticism, and there were some ideas concerning gravitation which he couldn't quite prove until the 1680's (see page 42).*

Newton's scheme can be described in two parts. The first part concerns the behavior of bodies with and without forces exerted on them (the three laws of motion). And the second part hypothesizes the existence of a force between any two bodies that have mass (the gravitational force). This chapter will concern itself with the three laws of motion, while the next chapter will concern itself with the gravitational force.

In Chapters 4 and 5, we will see how the three laws of motion and the

---

* For a detailed account of Newton's personality, see *Sir Isaac Newton*, E. N. da C. Andrade, Anchor Books, 1954.

gravitational force may be combined to explain the motion of falling bodies, the motion of the planets, the motion of man-made satellites, the tides, etc.

2-2 FIRST LAW OF MOTION

How does a body behave if there is no force acting on it?

Statement and Analysis

I. *If no net external force is exerted on a body, a body that is at rest tends to remain at rest, and a body that is moving tends to continue to move in a straight line forever.* (See Fig. 2-1.)

If no net external force is acting on a body,

I:
If the body is at rest, it will remain at rest forever.

II: ⟶ v
If the body is moving, it will continue to move in a straight line forever.

Fig. 2-1. Newton's first law of motion.

Newton's first law marks a radical departure from the older theories of motion. According to the older theories of motion, a body will remain at rest only if it is in its natural position. Also, according to the older theories, a body moves only to seek its natural position. The movement of a body towards its natural position is called natural motion and unless a body is undergoing natural motion, a moving force is required for it to continue moving. The only exceptions are the god-like bodies that lie in the heavens. These bodies only move in circles, because circles represent purity and perfection.

Newton, however, claims that if a body is at rest anywhere in the universe, it will remain at rest. Similarly, if a body is moving with a certain velocity, say $2 \measuredangle 0°$ m/sec, it will continue to move with that velocity forever. The concepts of natural position, natural motion, moving forces, and the idea that circles represent purity and perfection are eliminated in Newton's first law. In essence, Newton's first law says that unless an external force is applied to a body, the body will continue to do just what it is doing.

Why should a body behave like this? Because it has *inertia* or *mass*. The entire answer to this question is: If we say that matter is something that occupies space, then Newton defines mass as that *quality of matter* which causes it to obey the first law of motion. How much mass does a particular quantity of matter possess? The amount of mass that a particular quantity of matter possesses is related to Newton's second law of motion.

Newton's Laws of Motion

Comments

There are at least two subtle concepts associated with the first law. One is the concept of an inertial frame of reference, and the other is the concept of a system.

Concept of Inertial Frame of Reference

In the first law, what is meant by *"a straight line"*? That is, with respect to what frame of reference* does a body continue to move in a straight line?

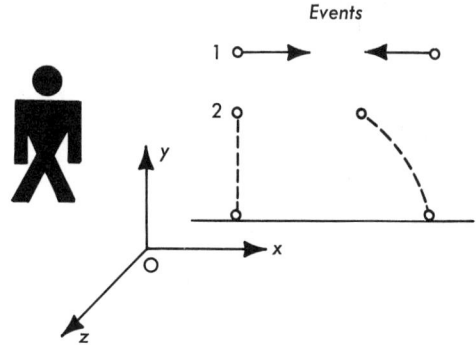

The man observes events by first chosing an origin (point O) and then measuring the location and characteristics of these events in terms of the x-, y-, and z-directions stemming from that origin.

Fig. 2-2. A frame of reference.

Newton felt that there existed an absolute space, and that a straight line was clearly defined relative to this absolute space. The substance comprising absolute space was called *ether*, and all of matter was pictured as being either at rest relative to this ether or in motion relative to this ether. (Just as fish exist in a sea of water, matter exists in a sea of ether.) In this way, the ideas of a straight line, absolute rest, and absolute motion came into being. Ether was said to represent an *absolute inertial frame of reference* or, more simply, an absolute inertial frame, and with respect to this absolute inertial frame, Newton's first law of motion is true. (In fact, all of Newton's laws are true.)

If Newton's laws were true only for the absolute inertial frame, they would be useless. However, Newton felt that a straight line in the absolute inertial frame appears as a straight line in any frame of reference moving at a constant velocity relative to this absolute inertial frame. So, Newton's laws are also true for *any* frame of reference moving at a constant velocity relative to the absolute inertial frame. These frames of reference are simply called *inertial frames*.

---

* The words "frame of reference" implies that an observer is observing events occurring in nature and is measuring the location and characteristics of these events relative to an origin convenient to him. See Fig. 2-2.

The earth was thought to be an inertial frame because it was thought to be moving at a constant velocity* relative to the absolute inertial frame. So Newton's laws must be true for an observer on the earth (see Fig. 2-3).

The only difficulty with Newton's scheme is that ether does not exist! This was accepted as fact when the relativity theories were developed in 1905†, and this discovery reopened the question "What is meant by *a straight line*?" Without ether, there is no absolute inertial frame, no absolute rest, and no absolute motion; in fact, nothing is absolute. (See Appendix 3 for a discussion of "straight lines.")

But there are inertial frames. Or, to put it another way, Newton's first law of motion is true for some frames of reference. In fact, it is true for an observer on the earth. So, in a very *ad hoc* manner, it was decided that if Newton's first law of motion is true for a particular frame of reference, that

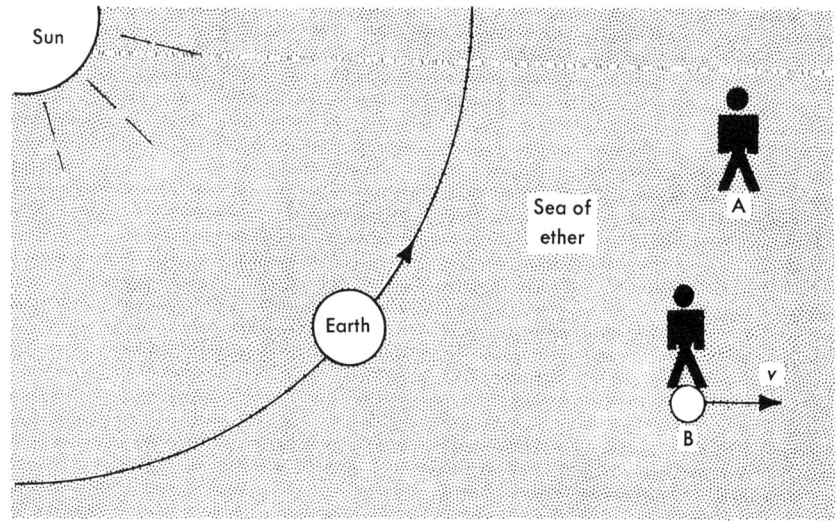

Man A is standing at rest in the ether and the whole universe moves within the stationary ether. Man A is in the absolute inertial frame and so man A can determine whether or not a body is in a state of absolute rest or absolute motion. Newton's first law is valid for man A. If man B is moving at a constant velocity relative to man A, Newton's first law is valid for man B also. Man B is in an inertial frame.

Fig. 2-3. **Ether, the absolute inertial frame and an inertial frame.**

---

* The earth is actually moving in an elliptic path about the sun and the sun is moving in an elliptic path in the Milky Way Galaxy and the Milky Way Galaxy is moving. . . .

† See Isaac Maleh, *Modern Physics,* Merrill Physical Science Series.

Newton's Laws of Motion

frame of reference is called an *inertial frame*. If Newton's first law of motion is not true for a particular frame of reference, that frame of reference is called a *non-inertial frame*.*

Examples of Inertial and Non-Inertial Frames

Let's consider some examples of inertial and non-inertial frames. Picture three men, one on a merry-go-round, one on a train moving at 60 mph, and the third standing on the ground.

If each of our three men took a ball and *placed* it on the floor of his reference frame, what would he observe? (Assume no friction.) See Fig. 2-4.

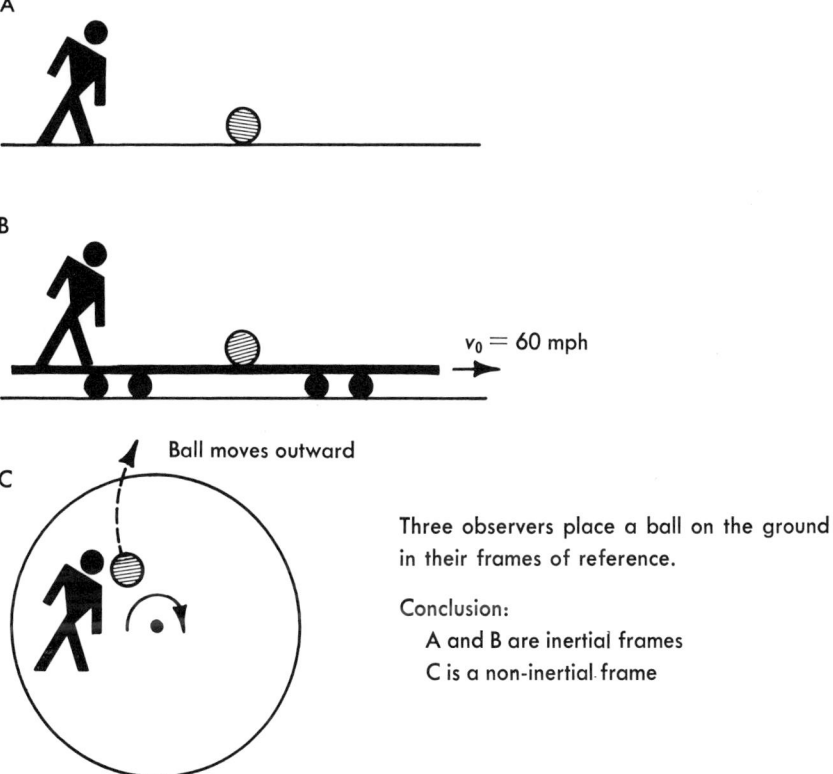

Three observers place a ball on the ground in their frames of reference.

Conclusion:
  A and B are inertial frames
  C is a non-inertial frame

Fig. 2-4. Test of inertial and non-inertial frames.

Merry-Go-Round	Train	Ground
The ball will move outward.	The ball will remain where it was placed.	The ball will remain where it was placed.

---

* It seems very peculiar to make a law and then say that if it is true the frame of reference is such and such and if it is not true the frame of reference is not such and such. Indeed, any law must be valid for some frame of reference. But the concept is successful, and it is difficult to argue with success.

If each of our three men took a ball and *rolled* it on the floor of his reference frame, what would he observe? (Assume no friction.) See Fig. 2-5.

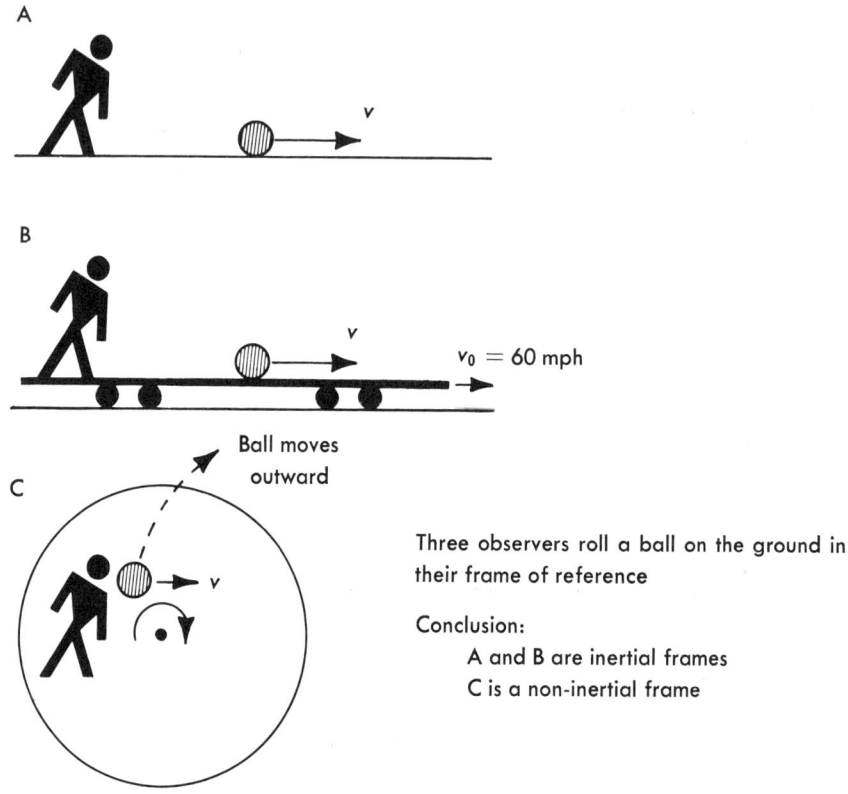

Three observers roll a ball on the ground in their frame of reference

Conclusion:
    A and B are inertial frames
    C is a non-inertial frame

Fig. 2-5. **Test of inertial and non-inertial frames.**

Merry-Go-Round	Train	Ground
The ball will move outward in addition to moving with its initial velocity.	The ball will continue to move with the same initial velocity.	The ball will continue to move with the same initial velocity.

Since the ball's behavior on the merry-go-round does not obey Newton's first law, the man on the merry-go-round is in a non-inertial frame. The ball's behavior on the train and on the ground does obey Newton's first law, so both the man on the train and the man on the ground are in inertial frames.

Newton's Laws of Motion

Concept of System

The second subtle concept in Newton's first law of motion involves the significance of the word *external*.

The first law of motion divides the universe into two systems: the body whose motion we are talking about is one system and everything external to that body is another system. Under this division, the body itself can do nothing to alter its motion; only external forces can alter the motion of the body.

Thus, we see that in the few words that constitute Newton's first law of motion lie many hidden concepts.

2-3 SECOND LAW OF MOTION

A baseball thrown horizontally will slow down; its velocity is changing because air is exerting a frictional force on the baseball. A bowling ball rolling down an infinitely long alley will eventually stop; the friction exerted by the floor of the alley, as well as air friction, has caused the velocity of the bowling ball to change to zero. If the air did not exist, and if the alley were frictionless, both the baseball and the bowling ball would continue to move forever in accordance with Newton's first law.

However, air does exist, friction does exist, and these forces do alter the motion of a body. How? The frictional forces acting on the baseball and on the bowling ball caused them to accelerate negatively; i.e., to decelerate. Newton's second law tells us exactly how forces alter the motion of a body.

Statement and Analysis

II. *If a net external force is exerted on a body, the body will accelerate in the direction of this net external force. The magnitude of the body's acceleration will be directly proportional to the magnitude of the force and inversely proportional to the mass of the body.* (See Fig. 2-6.)

Acceleration is a vector quantity and so the direction as well as the magnitude of the acceleration must be given. Fortunately, force is also a

If F is exerted on body 1, body 1 will accelerate in the direction of F with a magnitude such that $a = \frac{F}{m}$ or $F = ma$.

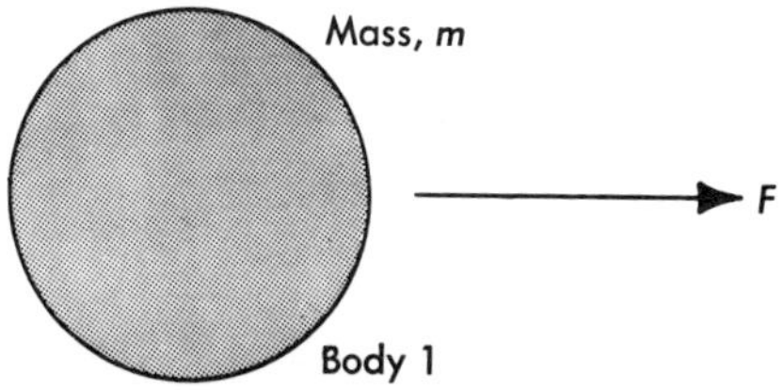

Fig. 2-6. Newton's second law of motion.

vector quantity, and so we can relate the direction and magnitude of the acceleration to the direction and magnitude of the force. So, the second law of motion says that if body A exerts a force on body B, body B will accelerate as follows:

1. *Direction of acceleration.* The direction of the acceleration will be the same as the direction of the force.

2. *Magnitude of acceleration.* The magnitude of the acceleration depends on the magnitude of the force, and on the *mass* of body B. If you keep the mass constant but increase the force, the acceleration increases, but if you keep the force constant and increase the mass, the acceleration decreases.

Recall that because a body has mass, it obeys Newton's first law of motion. But while a body will obey Newton's first law of motion *regardless* of the amount of mass it contains, the acceleration that a body acquires when a force is exerted on it depends on the amount of mass it contains.

The second law of motion is written as

$$[\text{Net, External force, On}] = \begin{bmatrix} \text{Mass} \\ \text{of} \\ \text{body} \end{bmatrix} \times \begin{bmatrix} \text{Acceleration} \\ \text{of} \\ \text{body} \end{bmatrix}$$

or

$$F = ma \qquad (2\text{-}1)$$

where

F is the net external force acting on the body
m is the mass of the body
a is the acceleration of the body.

Example 2-1:

A force of $20 \angle 0°$ newtons (nt) is applied to a body of mass 10 kg. What is the acceleration of the body?

$F = 20 \angle 0°$ nt

$m = 10$ kg $\rightarrow F$

$F = ma$

(The angle of zero degrees (i.e. $\angle 0°$) indicates that nothing happens in the vertical direction.)

$20 = 10a$

$a = 2 \, \dfrac{m}{\sec^2}$ to right

$a = 2 \angle 0° \, \dfrac{m}{\sec^2}$

A Unit of Force

How do we measure force? In the above example, I mention a force of 20 ∡ 0° newtons. What does that really mean? How much is one newton?

The definition of a newton (nt) of force depends on the choice of a unit of mass, the choice of a unit of length, and the choice of a unit of time. In the MKS system of units, a certain quantity of matter is called 1 kilogram, and if we exert enough force to accelerate this quantity of matter by 1 m/sec², we have exerted 1 nt of force. (See Figs. 2-7 and 2-8.) Mathematically,

$$F = ma$$
$$= 1 \text{ kg} \times 1 \frac{\text{m}}{\text{sec}^2}$$
$$= 1 \frac{\text{kg} \times \text{m}}{\text{sec}^2} = 1 \text{ nt}$$

Mass: Length: Time:

1 kg   1 m   1 sec

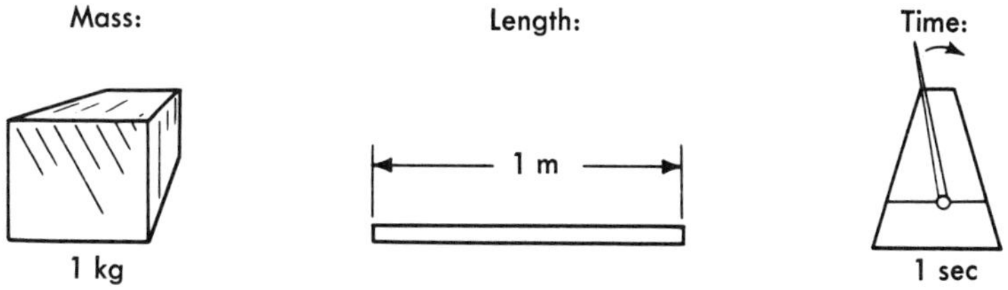

Fig. 2-7. Fundamental units of the MKS system (adopted by definition).

If $m = 1$ kg, and $a = 1 \frac{\text{m}}{\text{sec}^2}$,

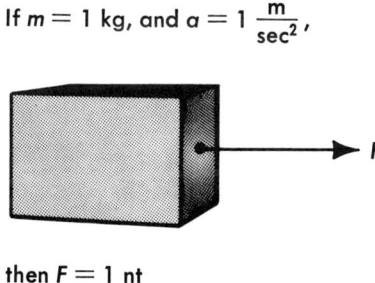

then $F = 1$ nt

Fig. 2-8. Definition of a unit of force.

Notice that force is just a composite of mass, length, and time. And in order to prevent writing out kg × m/sec² every time we have a force, as well as to honor Sir Isaac Newton, we call that composite of mass, length, and time a *newton*.

This definition is quite arbitrary, depending as it does on the units of mass, length, and time.* For example, in the English system of units, the unit of mass is a slug, the unit of distance is the foot, and the unit of time is the second. Using these units, the unit of force is a pound (lb). It turns out that approximately 4.55 nt = 1 lb. In this book we will almost always use the MKS system of units, so our unit of force is the newton.

Comments

Direction of Force, Direction of Velocity, and Direction of Acceleration

It is important to note that a force does not necessarily cause a body to *move* in the direction of the force—a force only causes a body to accelerate in the direction of the force. Now the direction of acceleration is not always the same as the direction of velocity. (See Fig. 2-9.)

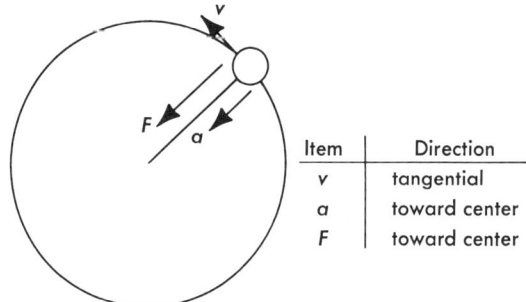

Fig. 2-9. Direction of F, a, and v for circular motion.

The Words "Net," "External," and "On"

There are at least three words to concern ourselves with in the second law: *net, external,* and *on.*

Net

In Example 2-1, a single force was being exerted on a body and we considered the motion of this body due to this single force. Suppose that a second force was also being exerted on the body—what happens then? The word *net* takes care of that situation. The two forces are added (vector addition), and the acceleration of the body is due to the net force applied.

External

The world is divided into two systems; the body is one system, and everything external to the body is another system. The acceleration of the body

---

* In the CGS system of units, the *gram* is the unit of mass, the *centimeter* is the unit of length, and the *second* is unit of time. If 1 gram is accelerated 1 cm/sec^2, the force exerted is called one *dyne*. One hundred thousand dynes is equal to one newton.

Newton's Laws of Motion

depends only on the external forces acting on the body; internal forces do not enter into the picture. (See the discussion on Newton's first law.)

On

The word *on* is a small, two-letter word, but it is very important. This word tells us that the acceleration of a body is due to the forces exerted *on* the body, and is not due to the forces exerted *by* the body. This fact will be very important when we discuss Newton's third law of motion.

2-4 THIRD LAW OF MOTION

A body will continue to move with constant velocity forever if no external forces are applied. If external forces are applied on the body, the body will change its velocity; i.e., it will accelerate. At first sight we seem to be in quite a dilemma, because everything depends on external forces. What can the body itself do to change its velocity?

Apparently nothing, that is until you consider Newton's third law. It is his most subtle law.

Statement and Analysis

III. *To every action there is an equal and opposite reaction, or if body A exerts a force on body B, then body B is simultaneously exerting a force on body A that is equal in magnitude, but opposite in direction.* (See Fig. 2-10.)

The subtlety in Newton's third law lies in the fact that if for every force there is an equal and opposite force, the net force will always be zero. Therefore, we can never have any change in velocity. We can never have any acceleration. But, obviously, bodies do accelerate. So if we blindly accept Newton's third law, we have a paradox.

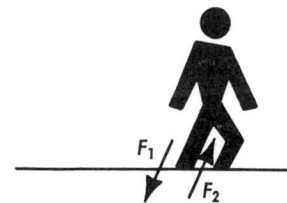

F_1 is the force exerted by the man on the floor,
F_2 is the force exerted by the floor on the man,
Motion of the man is due to F_2 (and any other forces acting <u>On</u> man).
Motion of the floor is due to F_1 (and any other forces acting <u>On</u> floor, i.e., walls, etc.).

Fig. 2-10. Action and reaction.

The solution to this paradox lies in the word *on* in Newton's second law. It is a small word, but it means a lot. The acceleration of a body is determined not by what the body is doing but by what others are doing to the body.

The reason a man can walk starting from a stationary position is not because he is exerting a force on the ground, but because the ground is

Example 2-2:

![10 nt arrow right, 10 nt arrow left, man pushing block]

While the man is exerting a force of 10 nt on the block, the block is exerting a force of 10 nt on the man. The acceleration of the block depends on the *net* force acting *on* the block. Thus the 10 nt exerted *by* the block on the man does not contribute to the motion of the block.

---

exerting a force on him. The man is clever enough to cause the ground to exert a force on him by first exerting a force on the ground; i.e., he knows that Newton's third law is true.

If a man jumps off a building there is nothing he can do to stop his fall unless he finds something on which to exert a force. That something will then return the force, via Newton's third law, and it is the return force *on* the man that will change his motion. (See Fig. 2-11.)

The first and second laws were discovered by Galileo and Descartes, but the third law is apparently original with Newton.

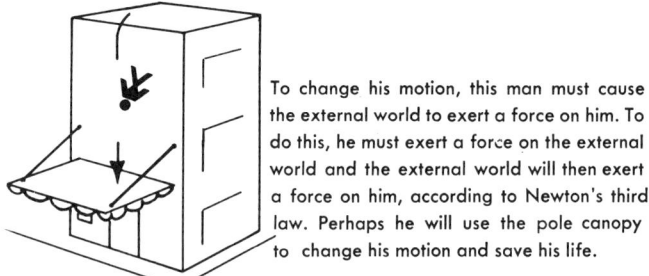

To change his motion, this man must cause the external world to exert a force on him. To do this, he must exert a force on the external world and the external world will then exert a force on him, according to Newton's third law. Perhaps he will use the pole canopy to change his motion and save his life.

Fig. 2-11. **The importance of Newton's third law.**

2-5 MOTION: ARISTOTELIAN VERSUS NEWTONIAN

The Newtonian concept of motion is quite different from the Aristotelian concept of motion.

According to the Aristotelian concept of motion, there are two different types of bodies—earthly bodies and heavenly bodies. Both of these types of bodies have a natural motion. Earthly bodies have a natural posi-

Newton's Laws of Motion

tion, so the natural motion of an earthly body is toward its natural position. Heavenly bodies have God-like souls, and such souls require a perfect form of motion, so the natural motion of a heavenly body is motion in a perfect circle or in a combination of perfect circles. Aristotle would claim that no external forces are required for a body to undergo natural motion, but any motion other than natural motion requires external forces. In a sense, Aristotle explained motion in terms of a destination and the attempt to reach that destination (earthly bodies), or in terms of an innate soul and the destiny of such a soul (heavenly bodies).

According to the Newtonian concept of motion, all bodies, earthly or heavenly, are equivalent. Also, there is no such thing as the natural position of a body; hence, a body cannot have a particular destination or goal. Also, heavenly bodies have no souls, and so the need for a perfect form of motion suited to the souls of heavenly bodies becomes unnecessary. The Newtonian concept of motion is embodied in Newton's three laws of motion. According to these laws, it is natural for a body to be either at rest or to be moving with constant velocity.* If an external force is exerted on a body, the body will change its velocity; i.e., the body will accelerate. Conversely, if a body is accelerating, an external force must be acting upon it.

Note that although Newton claims there is no such thing as the natural position of a body, he claims that there is such a thing as the natural *state* of a body; i.e., the body is either at rest or is moving with constant velocity. Also, Newton claims that any state other than this natural state requires an external force; e.g., circular motion would require an external force.

Newton's laws of motion are presumed to be applicable to all bodies, earthly or heavenly, animate or inanimate. Therefore, human beings as well as material bodies fall within the realm of Newton's laws of motion. However, human beings are clever. If a man wants to propel himself, he utilizes Newton's third law of motion or some conservation laws (see Chapter 6). Of course, Newton's third law of motion and the conservation laws are also applicable to material bodies, but these passively obey the laws, while human beings actively utilize them.

The Importance of Initial Conditions

Although Newton's laws of motion are the same for all bodies, the motion of all bodies is obviously not the same. Why?

Sometimes the net external forces acting on different bodies are different, and sometimes the masses of different bodies are different. Both of these factors can cause the acceleration, and hence the motion, of different bodies to be different. However, even if the same external force were exerted on each of two identical bodies, the motion of these two bodies may not be the same. Their acceleration would be the same, but their actual motion may not be the same. The actual motion of a body is determined by the initial position of the body and the initial velocity of the body *as well as* by the net

---

* Motion with constant velocity is called *uniform motion*.

external force and the mass of the body. (By definition, the initial position of a body and the initial velocity of a body are called the *initial conditions* of a body.) See Fig. 2-12. For example, in Sec. 1-2, we saw that, although the paths of the marbles were different, the same equation was used to describe their motion. (The equation is a mathematical representation of Newton's laws of motion as applied to bodies moving under the influence of gravity.) The actual path chosen by and predicted for each marble depended on the initial condition of each marble.

Once the initial conditions of a body, the mass of the body, and the net external force acting on the body are known, Newton's laws predict exactly what the future motion of this body will be. Newton's laws are said to be *deterministic*, because there is no ambiguity associated with their predictions.

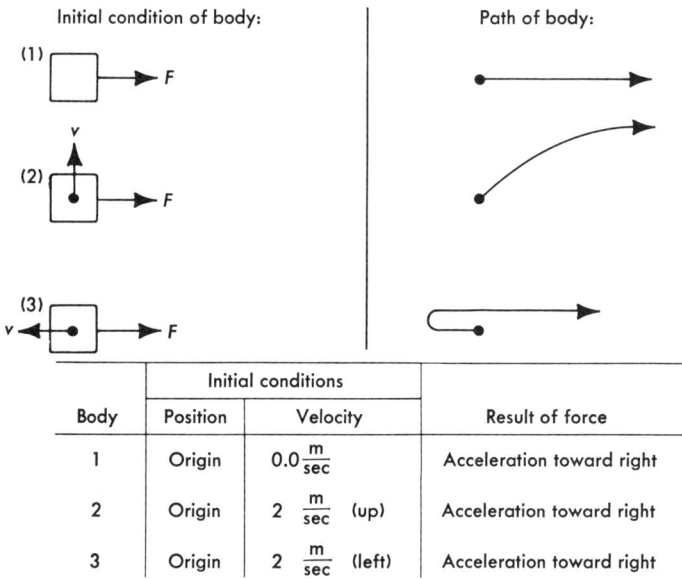

Fig. 2-12. The importance of initial conditions in the Newtonian scheme.

2-6 REVIEW

There are two major philosophies behind Newton's laws of motion.

The first major philosophy is the idea of a natural state of a body. A body is in its natural state when it is at rest or when it is moving with constant velocity. Any change in the velocity of a body is unnatural and requires an external force.

The first and second laws of motion incorporate this philosophy, and in so doing establish the following concepts and definitions:

Inertial frame	An inertial frame of reference is a frame of reference in which Newton's first law of motion is true.
Mass	Mass is that quality of matter which causes it to remain at rest if it is already at rest or to remain in motion if it is already in motion, unless there is a net external force acting on it. Because this definition of mass has to do with motion, it is more properly called *inertial mass*.
System	Since only external forces can cause a change in velocity, the world is divided into two systems: the body whose motion we are examining is one system, and everything else in the world is the external system. Only forces from the external system can cause a change in the velocity of the body.

What can a body, all by itself, do to change its velocity? According to the first and second laws, nothing. According to the third law, however, if the body can exert a force on the external system, the external system will then exert an equal and oppositely directed force on the body. This *reaction force* will cause a change in velocity of the body.

The second major philosophy lies in what might well be called the fate of moving bodies. In contrast to Aristotelian theory, which essentially tells us that bodies have either destinations or intrinsic destinies, and that they are moving in order to reach their destinations or in order to fulfill their destinies, Newton's laws tell us what a body will do provided that we know the initial conditions of the body, the mass of the body, and the net external force acting on the body. According to Newton's laws, there is no ultimate destination or intrinsic destiny for any particular body; rather, there is the present and according to the present, there is the future. The initial conditions of a body are very important in determining the future of the body.

How similar the situation is in a human life. Born a prince or a pauper, born an occidental or an oriental, born in good health or in poor, born under a lucky star or. . . . The initial conditions of a human being are extremely important in determining his future.

PROBLEMS

1. Define an inertial frame of reference.
2. If I were falling from a tall building, would it be possible for me to exert a force on myself to terminate my fall?
3. A sled is moving with a velocity of 5 m/sec due east. If the frictional force on the sled is 10 nt due west, what force must be exerted on the sled so that it maintains a velocity of 5 m/sec? (Give magnitude and direction.)
4. If a net external force of 8 ∡ 0° nt is exerted on a 2-kg body, what will be the resulting acceleration of the body?
5. If a net external force is exerted on a body, the body will accelerate according to its mass. Is the direction of the acceleration always the same as the direction of the force? Is the direction of the velocity always the same as the direction of the force? Is the direction of velocity always the same as the direction of acceleration? [*Hint:* Re-read Section 1-3.]
6. If I exert a force of 5 ∡ 0° nt on a book, the book will exert a force of 5 ∡ 180° on me. If the force that I exert on the book is equal but opposite to the force that the book exerts on me, why is it that the book accelerates but I don't? [*Hint:* Investigate the meaning of the words *net, external,* and *on* in all of Newton's laws.]

Chapter 3

Newton's Law of Gravitation

3-1 INTRODUCTION

It's not every day that the earth falls directly into the sun. In fact, it has never done so and probably never will. However, it happens practically every hour that a book, a pencil, or something similar falls to the ground here on earth. It's fantastic, but true, that the same type of force that causes the book to fall to the ground is being exerted on the earth by the sun, but the earth will never fall directly into the sun.

It took the genius of Sir Isaac Newton to relate the two forces, and to equate the laws of motion obeyed by both the earth and the falling book. We discussed Newton's laws of motion in an earlier chapter; we will discuss Newton's law of gravitation in this chapter.

3-2 THE GRAVITATIONAL FORCE

Background

Take an apple, raise it 5 or 6 feet above the ground, and let it go. You know, I know, and Newton knew what would happen. The apple will fall toward the earth. That's just the way things are. This phenomenon is so common that we hesitate even asking why. However, Newton and, I'm sure, many others, did ask why, and how, etc., etc. After all, there is nothing touching the apple.

Newton said that the apple doesn't just fall toward the earth, it accelerates toward the earth. And, if the apple is accelerating, there must be a force acting on it (see Newton's second law of motion). The obvious answer is that the earth is attracting the apple with a certain force. This is called the *gravitational force*.

But there is an unusual aspect to this gravitational force. If the apple were dropped from a tree-top or a mountain-top it would still accelerate toward the earth. Is there no limit to the influence of the earth on the apple? That is to say, "Is there no limit to the gravitational force?" Does it, perhaps, extend as far away as the moon?

Indeed, the moon is orbiting the earth in a nearly circular path, and we know that if a body is moving in a circle, it is accelerating toward the center of that circle.* Thus, the moon is accelerating toward the earth, and so the gravitational force does extend at least as far as the moon.

Does the gravitational force extend as far as the sun? Yes, and farther. We will soon see that the gravitational force extends infinitely far. (See Fig. 3-1.)

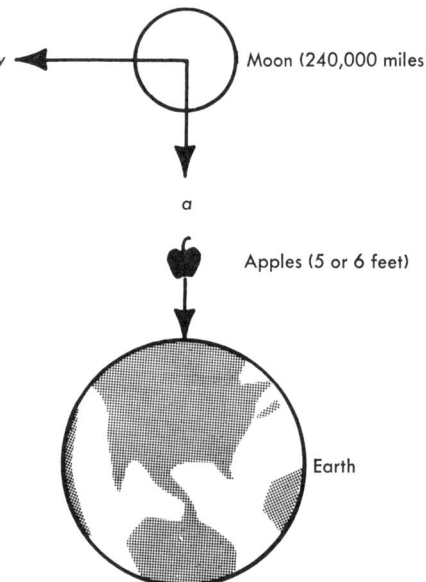

Fig. 3-1. **Range of the gravitational force.**

What does the gravitational force depend on? Is there anything peculiar about the earth, the apple, or the moon, or do all bodies exert gravitational forces on each other? We will soon see that producer of the gravitational force is the mass of a body.

Statement and Analysis

Newton hypothesized that every body has mass, and that between any two bodies there exists a so-called *gravitational force* which depends on the mass of each of the bodies and on the distance between the bodies.

---
* See Sec. 1-3.

Newton's Law of Gravitation

As shown in Fig. 3-2, the direction of the force exerted by body 1 on body 2 is toward body 1 and, symmetrically, the direction of the force exerted by body 2 on body 1 is toward body 2. These two forces are actually action-reaction pairs; i.e., because body 1 is exerting a force on body 2, body 2 is exerting an equal but oppositely directed force on body 1.

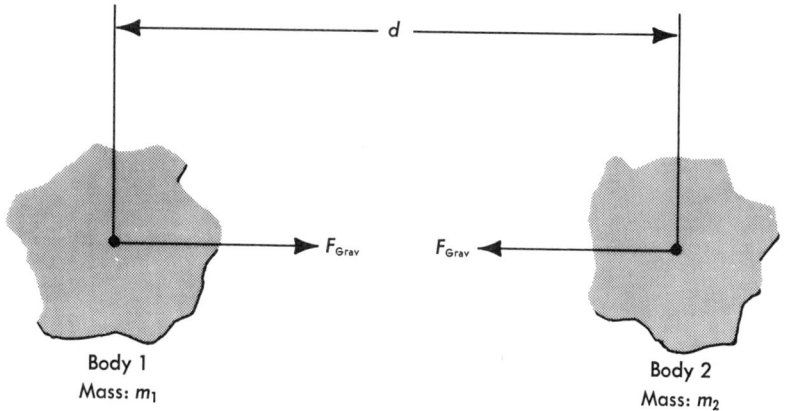

Fig. 3-2. The gravitational force.

Therefore, the reason the apple accelerated toward the earth is: since the apple has mass and the earth has mass, the earth exerted a force on the apple and the apple accelerated toward the earth.

"That's a fine chain of logic but according to the third law of motion, if the earth is exerting a force on the apple, the apple must also be exerting an equal force on the earth. I admit that this thought is unusual, but if we accept Newton's third law and if we have one force, we must have the reaction force also. Isn't that correct?"

"Yes."

"Now, since the apple is exerting a force on the earth, the earth must be accelerating toward the apple. But it doesn't, so Newton is all wrong. Too bad, Isaac, it was a good try; try again."

"But the earth *is* accelerating towards the apple, the amount of acceleration is just too small to detect."

("A likely story.")

The acceleration of a body depends on two things: (1) the force on the body, and (2) the mass of the body. While the force acting on the earth is the same as the force acting on the apple, the mass of the earth is a million-billion-billion times larger than the mass of the apple, so the acceleration of the earth is a million-billion-billion times smaller than the acceleration of the apple. Thus, although both the apple and the earth are accelerating toward each other, the acceleration of the apple is perceptible while the acceleration of the earth is not. (See Fig. 3-3.)

The magnitude of the gravitational force can be expressed mathematically as:

$$F_{\text{Grav}} = G \frac{m_1 m_2}{d^2} \tag{3-1}$$

where

F_{Grav} is the gravitational force
G is a universal gravitational constant $= 6.7 \times 10^{-11} \frac{\text{nt} \times \text{m}^2}{\text{kg}^2}$
m_1 is the mass of body 1
m_2 is the mass of body 2
d^2 is the square of the distance separating the two bodies.*

A gravitational force exists, and so acceleration will ensue:

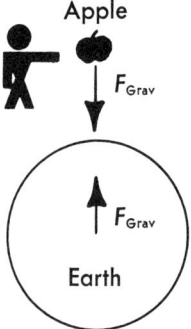

The acceleration of the apple depends on F_{Grav} and m_{apple}.
The acceleration of the earth depends on F_{Grav} and M_{earth}.
Since the m_{apple} is very much less than M_{earth}, the acceleration of the apple is very much more than the acceleration of the earth.

Fig. 3-3. **The earth and the apple.**

Example 3-1:

What is the gravitational force between two bodies ($m_1 = 1000$ kg, $m_2 = 3$ kg) that are 4 m apart?

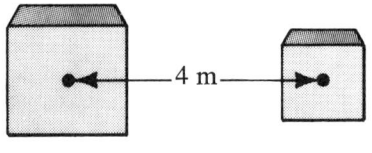

$m_1 = 1000$ kg, $m_2 = 3$ kg

$$F = G \frac{m_1 m_2}{d^2}$$

$$= 6.7 \times 10^{-11} \times \frac{1000 \times 3}{4^2}$$

$$= 6.7 \times 10^{-11} \times \frac{3000}{16}$$

$$= 1256 \times 10^{-11} \text{ nt}$$

$$F = 0.000\ 000\ 012\ 56 \text{ nt}$$

---

* The distance involved is the center-to-center distance. Although this seems very reasonable, it is not easy to prove. It is felt by many people that Newton delayed publication for many years until he could prove that the distance involved in Eq. (3-1) was the center-to-center distance.

G, The Universal Gravitational Constant

Why G?

Yes, why do we need G?

A word of explanation is necessary to understand the universal gravitational constant, G. As I mentioned earlier, the unit of mass was chosen to be 1 kilogram, the unit of distance to be 1 meter, and the unit of time to be 1 second. Finally, forces are measured in terms of the acceleration they impart on this unit of mass. In particular, the unit force is that force which will give 1 kg of mass and acceleration of 1 m/sec². For convenience, that unit of force is called a newton. Now if we place two 1-kg masses exactly 1 meter apart, what will be the gravitational force between them? Without knowing about G, we might say:

$$F_{\text{Grav}} = \frac{1 \text{ kg} \times 1 \text{ kg}}{1^2 \text{ m}^2} = \frac{1 \times 1}{1^2} \frac{\text{kg}^2}{\text{m}^2}$$
$$= 1 \text{ unit of force}$$

But we already have a unit of force—the newton. The question is, "Is the gravitational force between two 1-kg masses exactly 1 meter apart equal in magnitude to the force that we call one newton?"

The answer is *No!*

Therefore, if we insist on keeping the earlier definition of the newton which originated from Newton's second law of motion, we must insert a *proportionality factor*. The proportionality factor, G, relates the force due to gravitational forces to the force defined via Newton's second law of motion.

The numerical value of G can be determined by experiment. G is a very small number.

The Measurement of G

Henry Cavendish (1731-1810) was a very strange man. Although he lived almost 80 years, he probably uttered fewer words in his whole lifetime than most of us do in just 5 years.* Henry Cavendish measured the value of G.

The experimental measurement of G was very elusive because the gravitational force between two ordinary bodies is extremely small.† This fact caused many unforeseen problems. For example, the heat from Cavendish's body caused local temperature variations which in turn caused local air cur-

---

* Although he was of noble birth and inherited a fortune, he was quite frugal. Also he was so shy that he received few if any guests in his house; he ordered dinner by leaving a note on a hall table; he was either so morbidly afraid or so shy of women that he objected to any communication with his female domestic servants. Although he contributed enormously to the growth of science and was made immortal by the naming of a national laboratory in his honor (The Cavendish Laboratories in England); he himself seemed to miss the joy of life (See *Heroes of Science*, by W. Garnett, published in England about 1885).

† Of course, when huge bodies like the sun, moon, and earth are involved, the gravitational force becomes appreciable.

rents. These air currents disturbed the measurements. Cavendish finally had to enclose his apparatus and peer through a glass wall from another room in order to make his measurements. In spite of all the difficulties, in 1798, Cavendish succeeded in measuring the attraction between two known masses due to gravity alone. (See Fig. 3-4.) He found that, in the MKS system of units,

$$G = 6.7 \times 10^{-11} \frac{\text{nt} \times \text{m}^2}{\text{kg}^2}$$

The units associated with the proportionality factor were adapted in order to have the units of the gravitational force be newtons. Therefore, in the situation of two 1-kg masses exactly one meter apart, the gravitational force becomes

$$F_{\text{Grav}} = G \frac{m_1 m_2}{d^2}$$
$$= 6.7 \times 10^{-11} \times \frac{1 \times 1}{1^2}$$
$$F_{\text{Grav}} = 6.7 \times 10^{-11} \text{ nt}$$
$$= 0.000\,000\,000\,067 \text{ nt, or}$$
$$= 0.000\,000\,000\,015 \text{ lb}$$

Since G is so very small, it is only when the masses involved are very large that the gravitational force become appreciable. We will consider two examples of the gravitational force: one involving man and man, and the other involving the earth and man.

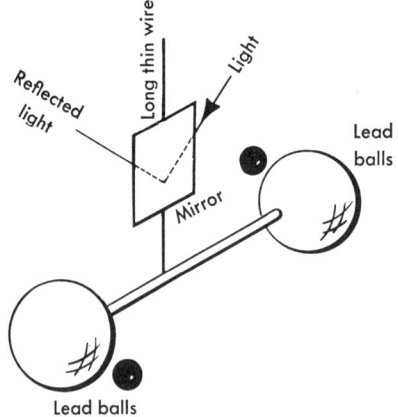

The attractive force between the lead balls twists the long thin wire, and the light reflecting off the mirror (which is attached to the wire) moves. The amount of twist depends on the amount of force, so by measuring the twist, we can measure the force between the lead balls. Since the masses and the distances involved are known, we can calculate G using Eq. (3-1).

Fig. 3-4. Measurement of G.

3-3 EXAMPLES OF THE GRAVITATIONAL FORCE

(1) What is the gravitational force between two men (each having a mass of 75 kg) standing 6.5 ft apart? (See Fig. 3-5.)

Newton's Law of Gravitation

$$F_{\text{Grav}} = G \frac{m_{\text{man}} m_{\text{man}}}{d^2}$$

where

$G = 6.7 \times 10^{-11}$
$m_{\text{man}} = 75 \text{ kg}$
$m_{\text{man}} = 75 \text{ kg}$
$d = 6.5 \text{ ft} = 2 \text{ m}$

so

$$F_{\text{Grav}} = 6.7 \times 10^{-11} \frac{75 \times 75}{2^2}$$
$$= 9.4 \times 10^{-8} \text{ nt (or } 2.1 \times 10^{-8} \text{ lb)}$$
$$= 0.000\ 000\ 094 \text{ nt (or } 0.000\ 000\ 021 \text{ lb)}$$

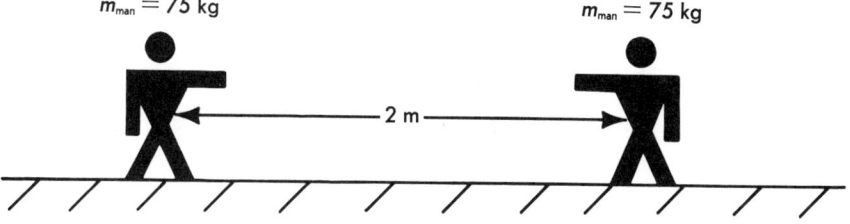

Note: A mass of 75 kg means that the man weighs 165 lb. (See Fig. 3-6.)

Fig. 3-5. The gravitational force between two men.

(2) What is the gravitational force between the earth and a man ($m_{\text{man}} = 75$ kg) standing on the earth? This gravitational force is usually called the weight of a man. (See Fig. 3-6.)

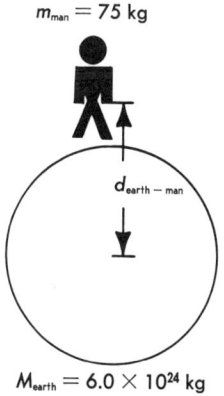

Fig. 3-6. The gravitational force between a man and the earth.

$$F_{Grav} = G \frac{M_{earth} m_{man}}{d^2_{earth-man}}$$

Before inserting numbers, notice that from Fig. 3-6, the distance between the center of the earth and the center of the man equals the radius of the earth plus about one and one-half meters. Since the radius of the earth is about 6.4 million meters (or 3,960 miles), the addition of one and one-half meters need not be considered. Thus, the gravitational force between the earth and the man is

$$F_{Grav} = G \frac{M_{earth} m_{man}}{R^2_{earth}}$$

where

$G = 6.7 \times 10^{-11}$
$M_{earth} = 6.0 \times 10^{24}$ kg
$m_{man} = 75$ kg
$R_{earth} = $ the radius of the earth (6.4×10^6 m).

Again, before inserting numbers, notice that because $d_{earth-man}$ can be replaced by R_{earth}, we can separate the gravitational force into two parts. One part is due to the earth and the other part is due to the man.

$$F_{Grav} = G \underbrace{\frac{M_{earth}}{R^2_{earth}}}_{\text{due to earth}} \times \underbrace{m_{man}}_{\text{due to man}}$$

$$= g \, m_{man}$$
$$= m_{man} \, g \qquad (3\text{-}2)$$

(Note that g is not identical to G.)
The value of g is

$$g = 6.7 \times 10^{-11} \frac{6.0 \times 10^{24}}{(6.4 \times 10^6)^2}$$

$$g = 9.8 \frac{m}{sec^2} \text{ (or, for convenience, } 10.0 \frac{m}{sec^2}\text{)}$$

Therefore, the gravitational force of the earth on the man, or the weight of the man, is

$$F_{Grav} = m_{man} \, g$$
$$= 75 \times 10.0$$
$$= 750 \text{ nt (or 165 lb)}$$

As long as the man is within 50,000 m (about 31 miles) of the surface of the earth, the approximation $d_{earth-man} = R_{earth}$ can be made, and Eq. (3-2) will be valid to within 2 per cent. Thus, the weight of a man can be determined by multiplying the mass of the man by g. In fact, the weight of any object

Newton's Law of Gravitation

within 50,000 m of the surface of the earth can be determined by multiplying the mass of that object by g.

3-4 SOURCE OF THE GRAVITATIONAL FORCE

Action-at-a-Distance or Gravitational Field

After I let go the apple in Fig. 3-3, there was nothing touching it. Yet the earth exerted a gravitational force on the apple and the apple obeyed, accelerating toward the earth. How does the earth exert a force on the apple? The earth has no hands, and no claws, so how does it produce a gravitational force on the apple?

The sun is 92.9 million miles away from the earth and there is nothing between us but empty space, yet the sun exerts a force on the earth and the earth obeys, accelerating toward the sun. How does the sun exert a force on the earth? The sun has no hands or claws either, so how does it produce a gravitational force almost a hundred million miles away?

Unlike most everyday forces, the gravitational force acts over empty space. It is not necessary for the masses involved to touch each other. It is very easy to say this, but why is it so? (See Fig. 3-7.)

One answer is, "That's the way gravitation is; it is an action-at-a-distance type of force." But that begs the question.

Another answer, slightly better, but not perfect, is, "The whole world is filled with a substance called 'ether,' and we are all living in a sea of ether. Any body that has mass produces disturbances in this ether, which we call a *gravitational field*. This gravitational field extends infinitely far from the originating body, and it can exert a force on any other body that has mass."

Thus, the earth disturbs the ether around it and the resulting gravitational field exerts a force on the apple. (The reaction force exerted by the apple on the earth is likewise caused by the disturbance that the apple produces on the ether surrounding it.) Similarly, the sun disturbs the ether around it and the resulting gravitational field exerts a force on the earth, almost a hundred million miles away.

Notice that the gravitational force is now a two-step process; a mass produces a gravitational field, and the gravitational field then exerts a force on other masses.

The ether that formed the gravitational field and exerted the gravitational force was thought to be the same ether which provided Newton with his absolute inertial frame of reference (see Sec. 2-2). Unfortunately, no one has ever detected this ether; it doesn't appear likely that it even exists.

So now we are back to the same old problem, "How does the gravitational force act over empty space?" Perhaps the best and most honest answer is, "At the present time, we don't really know."

One of the current ideas is that elementary particles called *gravitons* are constantly being emitted and absorbed by material bodies. The gravitons are the carriers of the gravitational force and essentially constitute the gravitational field. Offhand you might think that the word "ether" has merely been

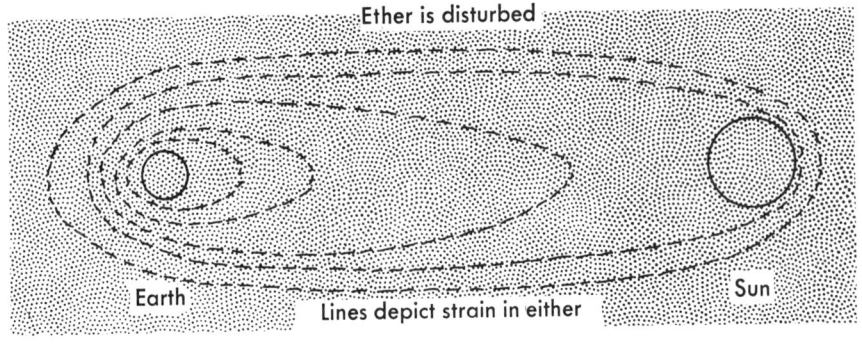
(a)

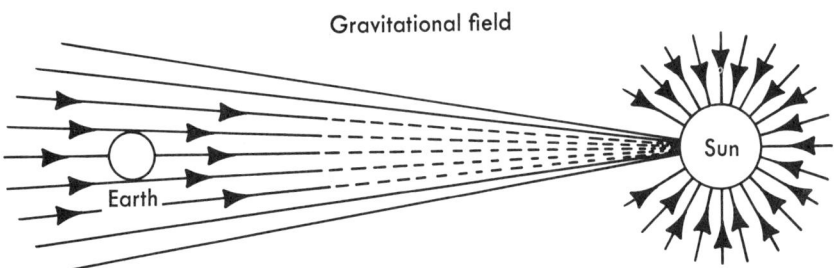
(b)

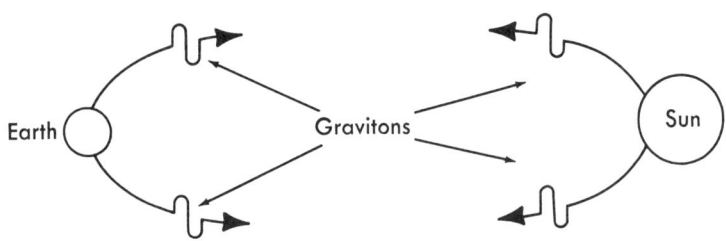
(c)

(d)

Fig. 3-7. The gravitational force, and its explanations over the years.

Newton's Law of Gravitation

changed to the word "gravitons," but there is a difference. While ether was supposed to exist independently of, and prior to, the existence of material bodies, gravitons are supposed to exist because of the existence of material bodies.* Gravitons do not exist independently of material bodies.

3-5 REVIEW

In order to understand one of the most common everyday occurrences (i.e., falling bodies), a gravitational force is hypothesized. In particular, all bodies are said to possess mass and because of their mass, all bodies exert forces on one another. This force is always attractive, and is termed the gravitational force. It can act over empty space (from the earth to the moon) and over enormous distances (hundreds of millions of miles). Other characteristics of the gravitational force are presented and discussed in Secs. 4-2, 4-3, 4-4, 5-2, 5-4, and 5-5.

The immediate problem is how to couple Newton's laws of motion and Newton's law of gravitation in order to understand motion in the solar system and motion on the earth. However, before coupling these two sets of laws, I want to emphasize some ideas presented in Sec. 2-3.

Because of the peculiarities of the laws of motion and the nature of velocity and acceleration, just because forces are exerted on bodies does not mean that the bodies get closer together. For example, it can be shown by Eq. (3-1) that the sun exerts a force of about 3.6×10^{22} nt on the earth—but the earth does not fall directly into the sun. We will see why this is so in the next chapter. For now, suffice it to say that there is no contradiction in Newton's scheme. It is internally consistent and closed.

PROBLEMS

1. An apple, mass 0.5 kg, is 5 meters above the surface of the earth. Compute the gravitational force that the earth exerts on the apple. What is the gravitational force that the apple exerts on the earth? (Use the data on page 46.)
2. The moon has a mass of 7.4×10^{22} kg and a radius of 1.7×10^6 meters. What is the gravitational force that the moon exerts on a 75-kg astronaut standing on the surface of the moon?
3. What is the gravitational force between a boy ($m = 75$ kg) and a girl ($m = 45$ kg) standing 3 m apart?
4. What is the gravitational force between the sun and the earth? ($M_{sun} = 2.0 \times 10^{30}$ kg, $M_{earth} = 6.0 \times 10^{24}$ kg, and $d_{sun-earth} = 1.5 \times 10^{11}$ m.)
5. What is the gravitational force between a man standing on the earth and the sun? (Let $m_{man} = 75$ kg.)

---

* Gravitons have yet to be detected in the laboratory, but the concept of a force field being produced by elementary particles has been verified in the lab. For example, the electromagnetic force field is caused by photons, and the nuclear force field is caused by mesons; and both photons and mesons have been detected in the lab.

Chapter 4

A Coupling of Newton's Laws of Motion and Newton's Law of Gravitation: I

4-1 INTRODUCTION

In Chapter 1 it was stated that any theory of motion had to explain the motion of all the planets in the solar system as well as the motion of all the bodies on the earth. In this chapter, I hope to show that Newton's laws of motion, coupled with Newton's law of gravitation, satisfies this demand.

4-2 THE PRINCIPLE OF EQUIVALENCE

Inertial Mass and Gravitational Mass

We have encountered the word "mass" twice so far; once when we discussed Newton's laws of motion, and once when we discussed Newton's law of gravitation. Each time the concept of mass was viewed as being responsible for a basically different characteristic of matter. So we really have two different kinds of mass, *inertial mass* and *gravitational mass*.

Coupling Newton's Laws of Motion and Law of Gravitation: I

Inertial mass	Gravitational mass
Because a body has mass, it obeys Newton's first law of motion. Also, from Newton's second law of motion, the mass of a body determines its acceleration under a given force: the greater the mass, the smaller the acceleration; the smaller the mass, the greater the acceleration.	Because a body has mass, it can exert a gravitational force on another body that has mass. The size of the force depends upon the mass of each body as well as on the distance between the two bodies.

The question is, "Given a material body, is the amount of inertial mass it possesses equal to the amount of gravitational mass it possesses?"

All experiments seem to say, "Yes!" Considering the different conceptual basis for each type of mass, this answer is unusual. We shall soon see how the principle postulated to explain this phenomenon, called the *principle of equivalence*, provides us with an understanding of why all bodies, regardless of weight, have the same acceleration toward the earth. Einstein used this principle extensively to develop his theory of general relativity. (see Fig. 4-1.)

Inertial mass = $m_{inertial}$
$m_{inertial}$ is the mass to be used in
the $F = ma$ equation.

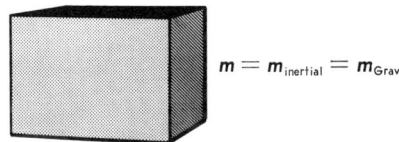

$m = m_{inertial} = m_{Grav}$

Gravitational mass = m_{Grav}
m_{Grav} is the mass to be used in
the $F = G\frac{mm}{d^2}$ equation.

Fig. 4-1. The principle of equivalence.

4-3 MOTION ON THE EARTH

On the earth we see falling bodies and bodies that are being pushed or pulled. An explanation of the behavior of falling bodies requires the use of Newton's laws of motion as well as the concept of the gravitational force. But bodies that are being pushed or pulled are being influenced by forces other than gravitational forces, and their behavior requires only the use of Newton's laws of motion.

Falling Bodies and the Gravitational Force

The main characteristic of falling bodies on the earth is that they all have the same acceleration. That is to say, both a ping-pong ball and a 5-ton truck fall toward the earth with an acceleration of 10.0 m/sec².* Therefore, if any two bodies have the same vertical history (see Sec. 1-2), they will both reach the ground simultaneously.

Why is this so?

Newton's laws say that a body will accelerate toward the earth because there is a gravitational force between the earth and the body, and secondly, that the amount of acceleration a body will experience is due to the mass of the body. Now, the gravitational force is related to the gravitational mass, and the acceleration is related to the inertial mass. But these two kinds of mass are exactly equal and they cancel each other's effect. Therefore, in a gravitational field, the mass of a body does not affect its acceleration and so, all falling bodies on the earth have the same acceleration.

Mathematically, the above paragraph is stated in the following way:

Gravitational force	**Second law of motion**
$F_{\text{Grav}} = G \dfrac{M_{\text{earth}} m}{R^2_{\text{earth}}}$	$F_{\text{Grav}} = ma$

Therefore,

$$G \frac{M_{\text{earth}} \cancel{m}}{R^2_{\text{earth}}} = \cancel{m} a$$

$$G \frac{M_{\text{earth}}}{R^2_{\text{earth}}} = a$$

or, from page 46,

$$a = 9.8 \text{ or } 10.0 \frac{\text{m}}{\text{sec}^2}$$

Motion Due to Forces Other Than Gravitational

If a body is at rest, it will remain at rest. However if you push it, the body will increase its velocity from zero to some value. As soon as you stop pushing the body, it will continue to move, but with a steadily decreasing velocity, until it is at rest again.

The force you exerted has accelerated the body, while air and surface friction have decelerated the body. If there were no friction, the body would continue to move forever at the velocity it acquired as a result of your pushing it. On earth, however, friction is always with us, and friction will tend to decelerate any moving body. The existence of friction has both clouded and delayed our understanding of motion.

Air and surface friction are quite substantial, and often, just to keep the

---

* We assume here that no air exists. Otherwise, air friction, winds, and buoyancy problems would affect the motion of bodies.

body moving at a constant velocity, a substantial force is needed to overcome these sources of friction. If the force exerted is too small, the unbalanced frictional force will decelerate the body and cause it to slow down. If the force exerted is too large, the unbalanced force will accelerate the body and cause it to move too rapidly.

4-4 MOTION IN THE SOLAR SYSTEM

The Solar System and Johannes Kepler

In 1609, after many years of tedious calculation, an Austrian by the name of Johannes Kepler presented two laws which described the motion of the planets in the solar system. In 1618, he presented a third law. These three laws became known as *Kepler's laws*.

The first of Kepler's laws describes the shape of a planetary orbit.

Kepler's First Law: All planets orbit the sun in an ellipse, with the sun at one of the focal points of the ellipse. (See Figs. 4-2 and 4-3.)

Circle:
$R + R =$ constant

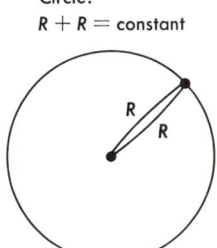

Distance from center to circumference back to the center is a constant (diameter).

Ellipse:
$a + b =$ constant

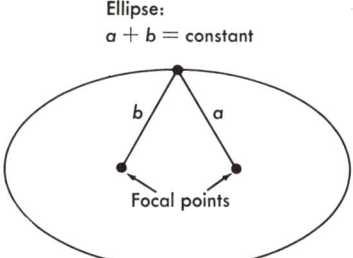

Distance from focal point to circumference back to focal point is a constant.

"An ellipse is a circle with two centers. Each center is called a focal point."

Fig. 4-2. Definition of an ellipse.

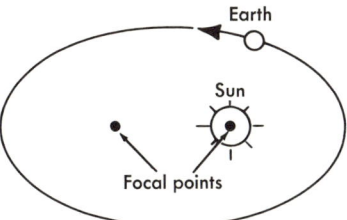

Elliptic orbit of the earth (exaggerated)

Fig. 4-3. Kepler's first law.

The second of Kepler's laws describes how the velocity of a planet changes as the planet orbits the sun.

Kepler's Second Law: The velocity of a planet is such that an imaginary line drawn from the sun to the planet sweeps out equal areas in equal times. (See Fig. 4-4.)

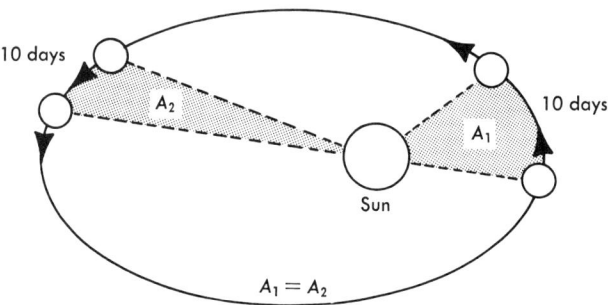

Fig. 4-4. Kepler's second law.

The third of Kepler's laws relates the average distance the planet lies from the sun to the time it takes the planet to revolve once about the sun (the time for one revolution is called a *period*).

Kepler's Third Law: The cube of the average distance a planet lies from the sun, divided by the square of the period of the planet, is a constant number. That is to say, if you take the cube of the average distance a planet lies from the sun and divide it by the square of the period of this planet, a certain number will result. If this is done for any other planet, the same number will result. (See Fig. 4-5.)

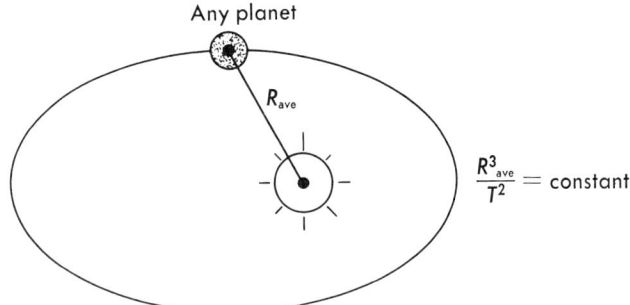

Fig. 4-5. Kepler's third law.

No rhyme or reason was given for Kepler's laws but they are the facts concerning the behavior of the planets in our solar system. Newton's laws predict the above facts, although, again, no rhyme or reason was given for Newton's laws, either.

The Solar System and Isaac Newton

How exactly did Newton's laws of motion and Newton's law of gravitation explain the behavior of the planets in the solar system? In order to appreciate the answer to this question, we must understand circular motion. The student should re-read Sec. 1-3. A summary of the basic ideas involved in circular motion follows.

Summary of Circular Motion

An object moving at constant speed in a circular orbit is accelerating. This acceleration is directed toward the center of the circle and has a magnitude of v^2/R, where v^2 is the square of the speed of the object and R is the radius of the orbit. (See Fig. 4-6.)

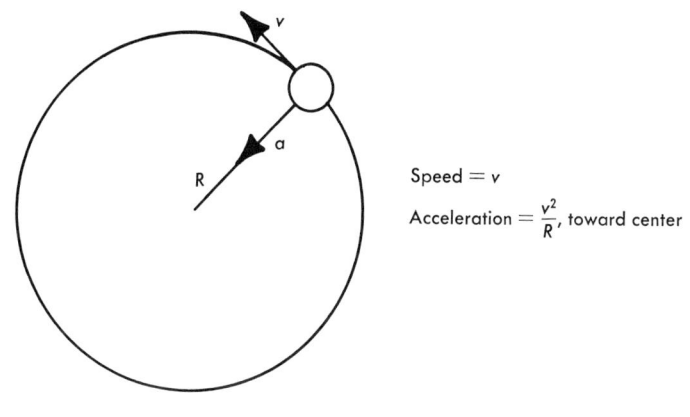

Speed $= v$

Acceleration $= \dfrac{v^2}{R}$, toward center

Fig. 4-6. Summary of circular motion.

Now, if we consider the earth as an object that is orbiting the sun in a perfect circle,* (radius, $R_{\text{sun}-\text{earth}}$; speed v), the earth is accelerating toward the sun with a magnitude $v^2/R_{\text{sun}-\text{earth}}$.

But according to Newton's second law of motion, if there is a force, there is an acceleration and, inversely, if there is an acceleration, there is a force; so, since the earth is accelerating, there must be a force on the earth. This force is the gravitational force between the sun and the earth.

Thus, in order to explain the motion of the earth about the sun, there is a coupling of Newton's laws of motion and Newton's law of gravitation. In fact the motion of all the planets about the sun are explicable via a coupling of the laws of motion and the law of gravitation. Let's see how this works out mathematically. (See Fig. 4-7.)

---

* The earth is actually moving in an elliptic path about the sun.

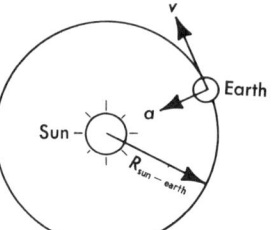

The sun has mass and the earth has mass; therefore, there is a gravitational force between them.

The earth is rotating about the sun in a circle (nearly) so the earth is accelerating toward the sun.

The gravitational force exerted on the earth by the sun causes the earth to accelerate toward the sun.

Fig. 4-7. Coupling of Newton's laws of motion and Newton's law of gravitation.

Gravitational force	Second law of motion
$F_{\text{sun-earth}} = G \dfrac{M_{\text{sun}} M_{\text{earth}}}{R^2_{\text{sun-earth}}}$ (3-1)	$F = M_{\text{earth}} a$ (2-1) $F_{\text{sun-earth}} = M_{\text{earth}} \dfrac{v^2}{R_{\text{sun-earth}}}$

$$G \frac{M_{\text{sun}} M_{\text{earth}}}{R^2_{\text{sun-earth}}} = M_{\text{earth}} \frac{v^2}{R_{\text{sun-earth}}}$$

$$G M_{\text{sun}} = \frac{v^2}{R_{\text{sun-earth}}} \times R^2_{\text{sun-earth}}$$

$$G M_{\text{sun}} = v^2 R_{\text{sun-earth}} \quad (4\text{-}1)$$

where

$F_{\text{sun-earth}}$ is the gravitational force between the sun and earth,
G is the universal gravitational constant,
M_{sun} is the mass of the sun,
M_{earth} is the mass of the earth,
v is the speed of the earth, and
$R_{\text{sun-earth}}$ is the distance from the sun to the earth.*

Eq. (4-1) states that the speed of the earth and the radius of its orbit is such that $v^2 R_{\text{sun-earth}}$ equals GM_{sun}.

Two comments should be made about Eq. (4-1). One concerns the importance of initial conditions on the actual motion of the earth, and the

---

* The distance is always taken to be the distance from the center of one body to the center of the other body. For circular orbits, this distance is the radius of the orbit; for elliptic orbits, this distance is the average radius of the orbit. See Fig. 4-5.

Coupling Newton's Laws of Motion and Law of Gravitation: I

other has to do with the equality of inertial and gravitational mass (i.e., the principle of equivalence).

Importance of Initial Conditions

As stated in Sec. 2-5, Newton's laws of motion, by themselves, do not tell us what the actual motion of a body will be. To determine the actual motion of a body, the initial position and the initial velocity of the body must be known. For example, if you simply drop this book from a ten-story window, with no initial velocity, it will plummet toward the earth. If, however, you throw it horizontally from a ten-story window, it will move in a curved path (a parabola) toward the earth. Given the initial conditions, Newton's laws would predict both of these phenomena. Newton's laws will tell us how long it takes the book to hit the ground, the velocity on impact, how far from the building it will land, etc.

Now about 4 or 5 billion years ago, the earth was placed 92.9 million miles from the sun and was given a certain velocity in a certain direction. With these initial conditions, Newton's laws predict that the earth will go around the sun in an elliptical path. (This is Kepler's first law of planetary motion.) Newton's laws also predict that an imaginary line between the sun and the earth will sweep out equal areas in equal times. (This is Kepler's second law of planetary motion.) And finally, Newton's laws predict that the cube of the average distance that the earth lies from the sun, divided by the square of the period, is a constant number. (This is Kepler's third law of planetary motion.)

But this behavior of the earth depends on the initial conditions of the earth. If, for some reason, the earth had simply been placed 92.9 million miles from the sun with no initial velocity in any direction, it would have plummeted toward the sun, just as the dropped book plummets toward the earth. (See Fig. 4-8.) If today, for some reason, the earth should stop moving,

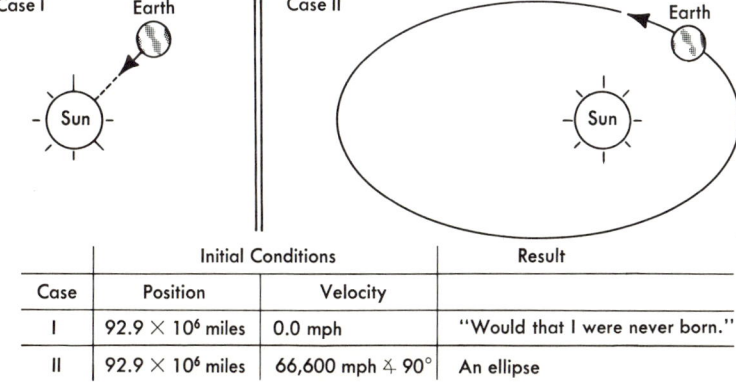

Fig. 4-8. **An event 4 or 5 billion years ago.**

	Initial Conditions		Result
Case	Position	Velocity	
I	92.9×10^6 miles	0.0 mph	"Would that I were never born."
II	92.9×10^6 miles	66,600 mph ∡ 90°	An ellipse

it would fall directly into the sun. Again, Newton's laws could predict the time it would take and the velocity on impact (if anyone would care to know). In the same way, each of the planets had an initial position (determined in some unknown way about 4 or 5 billion years ago) and an initial velocity relative to the sun. With these initial conditions, Newton's laws predict their future, forever and ever.

In a sense, Newton's laws predict a class of orbits; i.e., straight lines, ellipses, circles, parabolas, etc.* The particular one chosen by the body involved depends on the initial conditions of the body and the forces being exerted on the body.

Consequence of the Principle of Equivalence

What factors about the earth affect its motion about the sun? Does the mass, size, or shape of the earth have anything to do with its orbit about the sun? Do the oceans, mountains, or inner core of the earth have anything to do with its orbit about the sun?

The answer is "No!"

If you look again at Eq. (4-1), you'll notice that the mass of the earth cancels out of the relationship (this is due to the *principle of equivalence*). Notice also that GM_{sun} concerns the universal gravitational constant and the mass of the sun; it does not concern the earth at all.

The only items involved in the motion of the earth about the sun are the speed of the earth and its distance from the sun. The speed and distance are such that the above relationship is satisfied. If the distance between the sun and earth were less, and we still had a circular orbit, the speed of the earth would have to increase in order that $v^2 R_{sun-earth} = GM_{sun}$. Any object—a book, a candle, a human being, or a silver dollar—being 92.9 million miles away from the sun and having a speed of 66,600 mph in the same direction as the earth would orbit the sun once a year in an elliptic path. (See Fig. 4-9.)

4-5 THE MOON AND MAN-MADE SATELLITES

Just as the mass, size, or shape of the earth has nothing to do with its orbit about the sun (see Sec. 4-4), it can be shown that the mass, size, or shape of the moon has nothing to do with its orbit about the earth. The moon's speed and distance from the earth is such that

$$GM_{earth} = v^2 R_{earth-moon} \qquad (4\text{-}2)$$

---

* The mathematics associated with anything other than straight lines and circles is quite involved, so it must be taken on faith that elliptic orbits, parabolic orbits, hyperbolic orbits, etc., are indeed predicted by Newton's laws.

where
> M_{earth} is the mass of the earth,
> v is the speed of the moon, and
> $R_{earth-moon}$ is the distance from the earth to the moon.

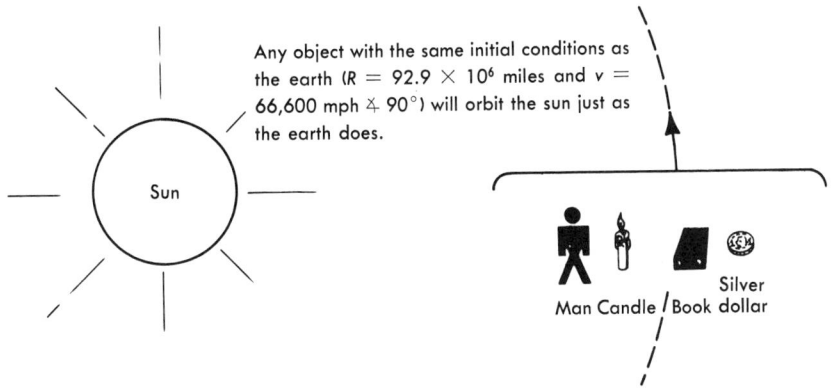

Fig. 4-9. **Consequence of the principle of equivalence.**

Any object 240,000 miles from the earth having the velocity that our moon has, would orbit the earth just as the moon does. All of the artificial satellites that orbit the earth have a speed and distance such that they satisfy

$$GM_{earth} = v^2 R_{earth-satellite} \qquad (4\text{-}3)$$

where
> v is the speed of the satellite, and
> $R_{earth-satellite}$ is the distance from the earth to the satellite.

At present, there are hundreds of artificial satellites orbiting the earth. They are at such distances and have such velocities that Eq. (4-3) holds true. Many of them will probably orbit the earth for millions of years. However, if for some reason they momentarily fail to satisfy Eq. (4-3) (i.e., their speed drops as a result of being hit by a meteor and their radius does not increase), they will begin to spiral toward the earth.

And land on top of our homes? No! Fortunately the atmosphere of the earth produces enough friction to melt most of the satellites before they reach the earth.

4-6 NEWTON'S LAWS AND KEPLER'S LAWS—A COMPARISON

Prior to Newton's approach to motion, Kepler's three laws of planetary motion explained the behavior of the planets. Now Newton's laws explain the behavior of the planets. Although, relative to the planets, both Newton's laws and Kepler's laws say the same thing, there is a basic difference between them.

While Kepler's laws concern only the planets, Newton's laws are not limited to the planets alone. For example, Kepler's laws describe the behavior of the satellites of the sun (i.e., the planets), but suppose that we were interested in the behavior of the satellites of the earth (i.e., the moon and rocket ships). Strictly speaking, Kepler's laws are not applicable, but Newton's laws are. Newton's laws predict that the behavior of the sun's satellites relative to the sun is the same as the behavior of the earth's satellites relative to the earth. That is to say, just as the earth orbits the sun in an ellipse with the sun at one focal point, the moon orbits the earth in an ellipse with the earth at one focal point.

Newton's laws are valid for motion in the solar system and for motion on the earth, and are valid for distances ranging from millions of miles to thousandths of an inch. Indeed, until the early 1900's it was thought that Newton's laws were valid for the motion of all bodies, under all circumstances.* (See Fig. 4-10.)

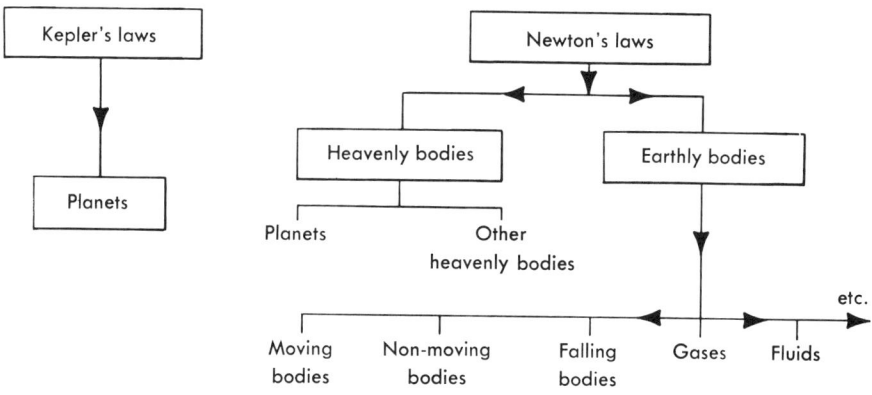

Fig. 4-10. Newton's laws and Kepler's laws, a comparison.

4-7 REVIEW

Newton's laws of motion describe the behavior of bodies with and without forces being exerted on them. The basic concept embodied in Newton's laws of motion is that motion with constant velocity is a natural state, and requires no external force for its continuance; it is change in velocity that is an unnatural state. Any change in velocity, therefore, requires an external force for its initiation and for its prolongation.

---

* In the early 1900's, it was realized that Newton's laws are really an approximation of what we feel today to be the laws of motion for all bodies. These new laws of motion are embodied in the quantum theory and the relativity theories. However, today, in the 1960's, some aspects of these new laws of motion are found wanting.

The actual path taken by a body when an external force is exerted upon it is determined by the magnitude and direction of the external force, by the mass of the body, and last but not least, by the initial conditions of the body. For hundreds of years the ever-present phenomenon of friction masked the relationship between velocity, change in velocity, and forces.

Together with his laws of motion, Newton had to introduce a so-called gravitational force. According to Newton's laws of motion, all bodies possess inertial mass. Now, according to Newton's laws of gravitation, all bodies also possess gravitational mass. Are gravitational mass and inertial mass identical? Yes! The principle of equivalence states that a body's inertial mass is exactly equal to its gravitational mass. By coupling these two sets of laws, and using the principle of equivalence, the motion of all the planets in the solar system, as well as the motion of all of the bodies here on earth, can be explained.

The gravitational force is a peculiar kind of force. It is the only force known to man wherein no quality of the body being influenced by the force can affect the behavior of the body. It is the initial conditions of a body that determine its behavior in a gravitational field. An example presented earlier is that all bodies on earth fall with the same acceleration; a ping-pong ball and a ten-ton truck will reach the ground simultaneously if their initial conditions are identical. No internal quality of the ping-pong ball or the ten-ton truck affects their motion. Similarly, the motions of the planets about the sun and the motion of the moon and the man-made satellites* about the earth are *not* determined by their mass, size, or shape, but by their initial conditions. This feature of the gravitational force derives from the equality of gravitational and inertial mass.

In spite of the success of Newton's laws, they have their weaknesses. A very glaring weakness is the inability to solve for the motion of three bodies which are exerting forces on each other simultaneously. A similar inability exists when four, five, etc. bodies are involved. Newton's laws only lead to exact answers when one body, two bodies, or an infinite number of bodies are involved. If more than two but less than an infinite number of bodies are involved, approximation techniques (see Sec. 5-4) or various conservation laws (see Chapter 6) must be used.

PROBLEMS

1. Using Eqs. (2-1) and (3-2), show that the acceleration of a 75-kg man who jumps off a cliff is 9.8 m/sec^2.
2. If the gravitational force that the sun exerts on the earth is 3.6×10^{22} nt, and the mass of the earth is 6.0×10^{24} kg, what is the resulting acceleration of the earth?
3. Are Newton's laws contained within Kepler's laws or are Kepler's laws contained within Newton's laws?
4. How does an ellipse differ from a circle?

---

* I am assuming that the rockets of these man-made satellites are turned off.

5. Due to the elliptic orbit of the earth, its closest approach to the sun is 91.4 million miles, while its farthest distance is 94.4 million miles. Using Kepler's second law, would you say that the earth is moving faster when it is 91.4 million miles from the sun than when it is 94.4 million miles from the sun, or vice versa?
6. What is the principle of equivalence?
7. In the equation describing Newton's second law of motion [Eq. (2-1)], what sort of mass is Newton talking about?
8. In the equation describing Newton's law of gravitation [Eq. (3-1)], what sort of mass is Newton talking about?

Chapter 5

A Coupling of Newton's Laws of Motion and Newton's Law of Gravitation: II

5-1 INTRODUCTION

This chapter is a continuation of Chapter 4. It presents other aspects of motion and gravitation, as well as some oddities of the law of gravitation.

5-2 WEIGHT AND WEIGHTLESSNESS

The Sensation of "Free Fall"

Let us consider the question of weight and weightlessness.

If one of the satellites circling the earth were a man, would he be weightless? No. Weight is the gravitational force of the earth on the object, and the gravitational force never goes to zero. To be sure, the force becomes smaller as the distance between the bodies increases but it never becomes zero.

Although newspapers and magazines are filled with words like "weightlessness," "no gravity," etc., the fact is that the gravitational force never dies, it just decreases as $1/d^2$. (See Fig. 5-1.)

We will compute the gravitational force between a man and the earth for two different situations. In both cases, the force is something other than zero. We will see that the feeling of weightlessness does not come about because the gravitational force is zero but rather because of the man's reaction to that force.

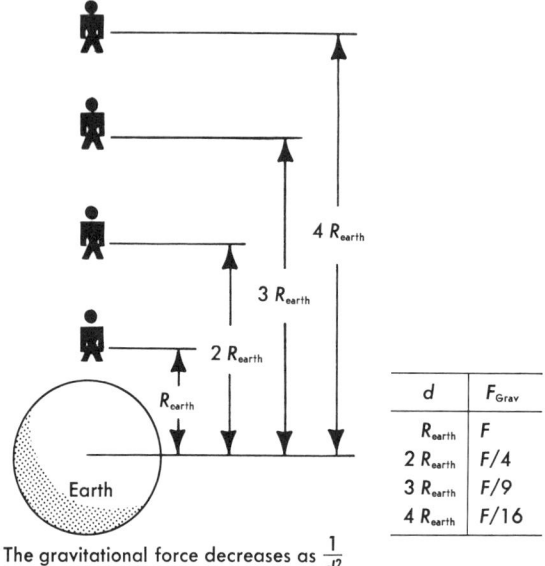

The gravitational force decreases as $\frac{1}{d^2}$

Fig. 5-1. **The infinite range of the gravitational force.**

From Sec. 3-3, we see that the force exerted by the earth on a 75-kg man standing on the earth is

$$F_{\text{Grav}} = m_{\text{man}}g \qquad (3\text{-}2)$$
$$= 75 \times 10.0$$
$$= 750 \text{ nt (or 165 lb)}$$

This force is the weight of the man. If the man were not standing on the ground, this 750-nt force would cause him to accelerate toward the center of the earth. But the man is standing on the ground, and the ground is exerting a force of 750 nt upward, thus negating the gravitational force due to the earth.

Now raise this man 20 m off the ground and drop him. He will accelerate toward the earth because there is nothing to counter the 750 nt* force. During the time that this poor man is accelerating freely toward the earth, he is said to be in "free fall," and he experiences the sensation of weightlessness.

Suppose now that this man is more than 50,000 m above the surface of the earth. For example, suppose that he is 4,000,000 m (or about 250 miles) above the surface of the earth. Now, the formal equation for the gravitational force [Eq. (3-1)] must be used, because there is no way of separating the force into a portion due to the earth and a portion due to the man.

---

* Twenty meters is still small compared to the radius of the earth, so we can still use $F_{\text{Grav}} = mg$. As noted at the bottom of page 46, we can use $F_{\text{Grav}} = mg$ until the object is 50,000 m or more from the surface of the earth.

$$F_{\text{earth}-\text{man}} = G \frac{M_{\text{earth}} m_{\text{man}}}{d^2_{\text{earth}-\text{man}}} \quad (3\text{-}1)$$

In this case, the distance between the center of the earth and the center of the man is 6,400,000 + 4,000,000, or 10.4×10^6 m. Using the appropriate numbers, we obtain

$$F_{\text{earth}-\text{man}} = 6.7 \times 10^{-11} \times \frac{6.0 \times 10^{24} \times 75}{(10.4 \times 10^6)^2}$$

$$= 6.7 \times 10^{-11} \times \frac{450 \times 10^{24}}{1.08 \times 10^{14}}$$

$$= 279 \text{ nt (or 61.4 lb)}$$

Note that the mass of the man (75 kg) remains the same, even though the gravitational force on the man decreases as the distance between him and the earth increases. Thus, the earth exerts 750 nt on the man when he is on the surface of the earth, and 279 nt when he is four million meters above the surface of the earth. Since there is a net force acting on the man, he must, according to Newton's second law of motion, accelerate toward the earth. (See Fig. 5-2.)

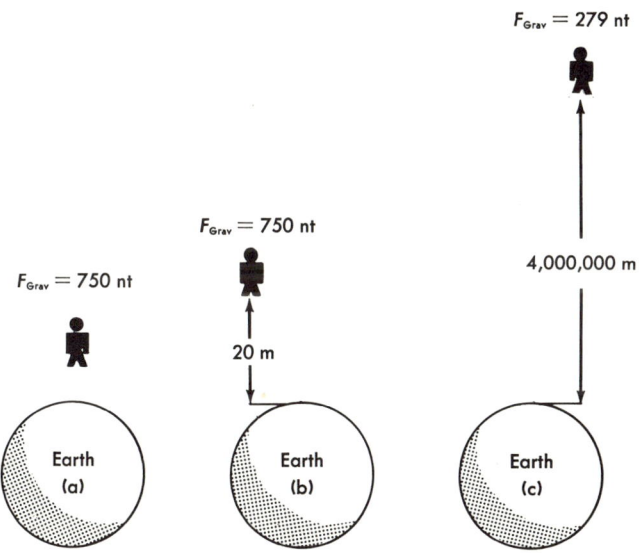

Fig. 5-2. The gravitational force on a 75-kg man at various distances from the earth.

Exactly how will this man accelerate? If our 75-kg man were simply placed four million meters above the surface of the earth, he would plummet toward the earth and arrive with disastrous results [Fig. 5-3(a)].

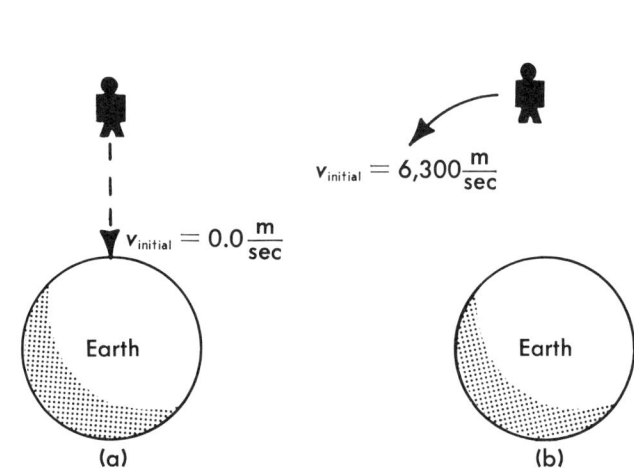

Fig. 5-3. **Two men in free fall: the sensation of weightlessness.**

However, if our 75-kg man had a particular initial velocity at this location, he would travel as shown in Fig. 5-3(b). In the first case, the man is accelerating toward the earth and moving toward the earth. In the second case, he is also accelerating toward the earth, but because of his initial conditions, he will orbit the earth. The magnitude of his acceleration is v^2/R and the direction is toward the earth. In either case, during the time interval in which our man is accelerating freely toward the earth, he is in "free fall" and experiences the sensation of weightlessness. In one case this time interval is short, while in the second case, the time interval is quite long (an eternity, perhaps?). It was the initial conditions that caused temporary weightlessness and disaster in one case, and eternal weightlessness in the second case.

Let's examine the characteristics of this sensation.

(1) A man in free fall cannot control his motion in the usual way.

Suppose that an astronaut, in free fall in outer space, wants to turn around, walk a little, or in some other way alter his motion. On the earth, he can do this by simply finding some body on which to exert the appropriate force. The appropriate reaction force will then be exerted on him (according to Newton's third law), and this reaction force will cause his motion to be altered. For example, if he wants to walk, he presses downwards and backwards onto the floor, and the floor then exerts an upward and forward force on him. The upward force supports his weight and the forward force causes him to move forward. However there are no floors in outer space. There are no floors, no walls, no shelves, or anything else. This feeling of partial helplessness promotes the sensation of weightlessness.*

(2) As a result of the principle of equivalence, no internal quality of a body affects its acceleration under a gravitational force (See Sec. 4-4). The inverse of this is "the acceleration of a body under a gravitational force does not affect any internal quality of the body."

Thus, when a body is accelerating under the influence of a gravitational force; i.e., when a body is in free fall, no internal quality of the body is affected. Because no internal quality is affected, a body cannot tell if it is in "free fall" or not. A man diving into a swimming pool (Fig. 5-4) is accelerat-

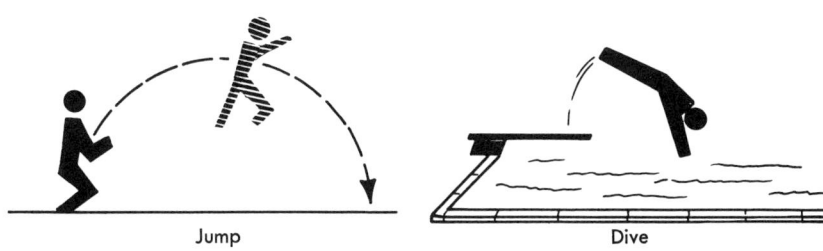

Fig. 5-4. **Other examples of temporary weightlessness.**

ing toward the earth from the moment he leaves the diving board until the moment he reaches the cool, blue water. His initial position and his initial velocity determine the shape of his dive. He, just like our astronaut, is in free fall during this time interval and he experiences the sensation of weightlessness from the moment he leaves the diving board until the moment he reaches the water. Similarly, when we jump off the ground, from the moment we rise off the ground until the moment we hit the ground again there is a gravitational force acting on us and we are accelerating freely toward the earth. We, just like our astronaut and our diver, are in free fall during this short time interval.

As shown in Problem 5 of Chapter 3, the sun is exerting a force of 0.45 nt on an average man here on earth. But we don't feel this force because the earth, with us on it, is rotating about the sun with the appropriate acceleration. We are in "free fall" relative to the sun's gravitational force, and so we experience the sensation of weightlessness as far as the sun's gravitational force is concerned. Although the earth is exerting a gravitational force on us,

---

* The space man can move by taking advantage of certain conservation laws to be discussed in Chapter 6. Without these, he would be quite helpless. We are neglecting the existence of air.

68   *Mechanics, Heat, and Sound*

the ground exerts an equilibrating force which prevents us from accelerating. We can feel the equilibrating force that the ground exerts on us. In a sense, it is the equilibrating force that makes us aware of the gravitational force.

5-3 TIDES

Just as the earth exerts a gravitational force on the moon, the moon exerts a gravitational force on the earth. However, the earth is not like the moon; the moon is a solid throughout, while the earth has a solid portion and a liquid portion. In fact, about 70 per cent of the earth's surface is water with an average depth of $2\frac{1}{2}$ miles.

Since there is a rigid connection between the various portions of a solid, the solid portion of the earth must act as one unit. However, there is no rigid connection between the various portions of a liquid, nor is there a rigid connection between the solid portion of the earth and the liquid portion of the earth. Therefore, the liquid portion of the earth can flow over, around, and about the solid portion of the earth. Tides are caused by this characteristic of liquids, by the change in strength of the gravitational force due to the moon as we go from one side of the earth to the other, and by the mutual rotation of the earth and the moon about a common center. First we will consider the situation without the mutual rotation and then we will add the effect of this mutual rotation.

Without Mutual Rotation (See Fig. 5-5a)

The earth as a solid mass with a liquid mass envelope.

(a) Without mutual rotation:

The gravitational force of the moon is 7% greater on A than on C, so the liquid becomes shaped like a teardrop.

(b) With mutual rotation:

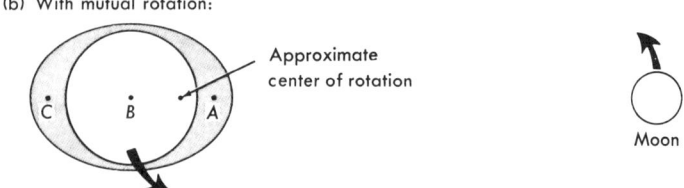

When the moon and the earth rotate about each other, the teardrop becomes an ellipsoid.

Fig. 5-5.   **The cause of tides.**

Coupling Newton's Laws of Motion and Law of Gravitation: II

For simplicity, let's consider this planet to be made of a solid, rigid mass completely covered with a liquid mass envelope. The mass at point A is 4000 miles closer to the moon than the mass at point B, and the mass at point B is 4000 miles closer to the moon than the mass at point C. Therefore, a stronger gravitational force will be exerted on A than on B and a stronger gravitational force will be exerted on B than on C. The greater the force the greater is the attraction, so that A will be attracted most strongly toward the moon, C will be attracted least strongly, and B will be somewhere in between. The shape of the solid portion will remain essentially the same; however, because of the "flowability" of liquids, the liquid portion of the earth will become like a teardrop. The shape of the teardrop is determined by the location of the moon; i.e., the cusp-like end points toward the moon.

With Mutual Rotation (See Fig. 5-5b)

Whenever a body is rotating about a point, every portion of the body is rotating about that point and so every portion of the body is accelerating toward that point. (See Circular Motion, Sec. 1-3.) In order for every portion of a body to accelerate toward that point, which is the center of rotation, an adequate force must be exerted. (This force is called a *centripetal force* because it is directed toward the center of rotation.) If no adequate force is available, every portion of the body will not accelerate properly and will tend to pull away from the center of rotation. This pulling away from the center of rotation is said to be due to a hypothetical *centrifugal force*. (Centrifugal means away from the center of rotation.) The magnitude of the centrifugal force increases with the distance from the center of rotation.

The teardrop becomes an ellipsoid because of the mutual rotation of the earth and the moon about a common center. The common center of rotation of this system is on a line between the earth and the moon, and is about 3000 miles from the center of the earth. Our teardrop is rotating about this point and so both the cusp-like portion and the flattened portion feel a centrifugal force. However, the cusp-like portion, being closer to the center of rotation, feels a smaller centrifugal force then the flattened portion. But, if we recall, the cusp-like portion had a larger gravitational force exerted upon it than did the flattened portion. It turns out that the *sum* of the smaller centrifugal force plus the larger gravitational force on the cusp-like portion of our teardrop *almost exactly equals* the *sum* of the larger centrifugal force plus the smaller gravitational force on the flattened portion of our teardrop. The result is that the flattened portion will elongate just as much as the cusp-like portion has already elongated and we will get an ellipsoid. One hump of our ellipsoid points toward the moon and one hump points away from the moon.

This picture of the earth, moon, and tides is slightly more complicated because of the presence of the sun. When the sun is along the line connecting the moon and the earth, the sun enhances the tidal force of the moon and we get the unusually high *spring tides*. However, when the sun is perpendicular to the line connecting the moon and the earth, the sun detracts from the tidal force of the moon and we get the smaller *neap tides*. (See Fig. 5-6.)

The effect of the sun on tides is not as large as that of the moon because the sun is 93,000,000 miles away, and 93,000,000 plus or minus 4000 is still essentially 93,000,000 miles. The moon, however, is only 240,000 miles away and 240,000 plus or minus 4000 may be important.* Indeed, the existence of tides shows it to be.

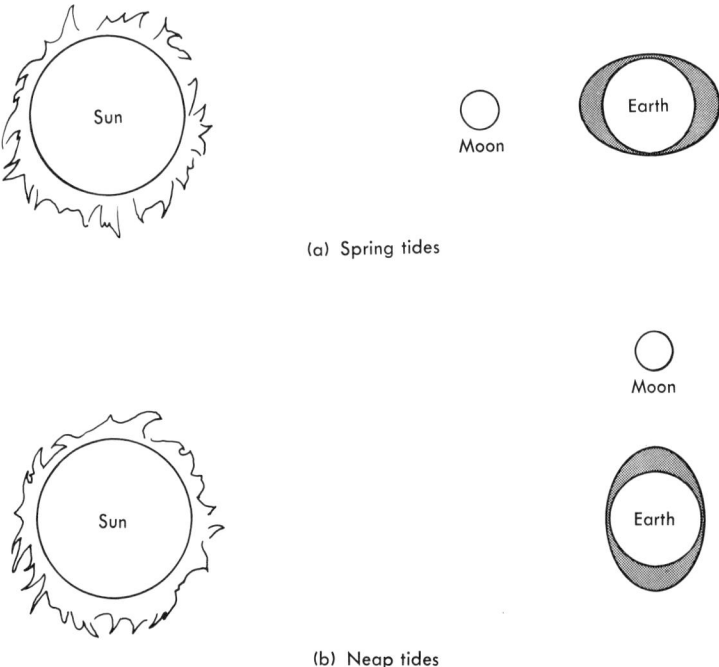

Fig. 5-6. Spring tides and neap tides. The influence of the sun on tides.

5-4 PERTURBATIONS OF THE PLANETARY ORBITS

The gravitational force exerted on the earth by the sun causes the earth to accelerate toward the sun. However, because of the initial conditions of the earth, the earth rotates about the sun in an elliptic orbit rather than plummeting directly into the sun. But the law of gravitation predicts that an attractive force exists between any two bodies possessing mass, and there are eight other planets, plus the moon.

Don't these planets as well as the moon attract the earth and essentially

---

* As we go from point A (Fig. 5-5) to point C, the gravitational force due to the moon changes by about 7%, while the gravitational force due to the sun changes by less than 0.1%.

interfere with the pull of the sun? Yes! The moon and all of the other planets perturb the orbit of the earth in its path about the sun. This perturbation is extremely small, however, because the combined mass of the planets and moon* is so much smaller than the mass of the sun. In fact, the total mass of all the planets and the moon is less than 1 per cent of the mass of the sun.

Although this correction factor is small, it can be computed, and the path of the earth is modified slightly. To find the exact motion of the earth, we first compute the major effect; i.e., that due to the sun, and then we add to it the perturbation caused by the planets, the moon, etc. Similarly, the exact motion of *any* planet in the solar system is computed by considering, first and foremost, the attractive force due to the sun, and, secondly, the perturbation caused by the other planets, the moon, etc. (See Fig. 5-7.)

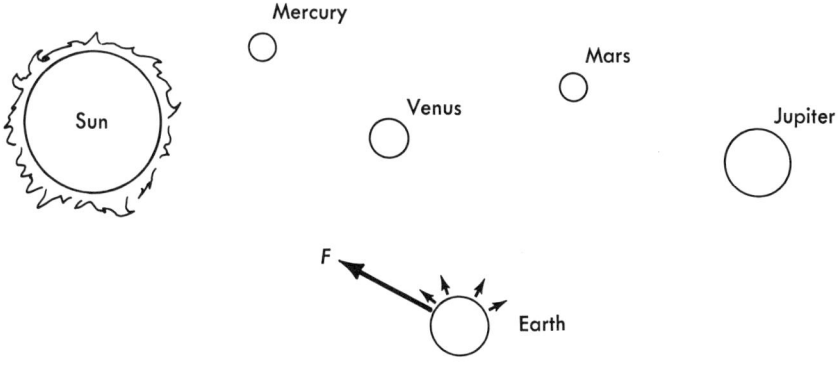

Since the net force on the earth is more than 99.99% due to the sun and less than 0.01% due to the planets, we can neglect the planets as a first approximation.

Fig. 5-7. Perturbation theory.

This technique of first taking into account the major effect, and then adding minor corrections, is called the *perturbation theory*. It is used because it is simply impossible to include all of the effects simultaneously. Fortunately, in our solar system there is one major body, the sun, which produces the major effect, and so the technique can be used. If the mass of the sun were about the same as the mass of the planets, the sun's effect would not be major, the perturbations would not be minor, and all effects would have to be considered

---

* However, the moon does affect the tides. This is not because the gravitational force is so large, but because it changes by 7% as one goes from the side of the earth nearest the moon to the side of the earth farthest from the moon.

simultaneously. This is an impossible task today, even with all our high-speed computers. In fact, one of the major problems in trying to understand the nucleus lies in the fact that all bodies in the nucleus are more or less equally important and, hence, perturbation theory is not applicable.

Perturbation theory and perturbation effects in the solar system have led to two major discoveries.

The Discoveries of Neptune and Pluto

By the turn of the nineteenth century, a total of seven planets were known and were being studied. Using perturbation theory, the orbit of the outermost planet, Uranus, was calculated as accurately as possible by the best scientific minds of the time. Scientists knew exactly where Uranus should be. Unfortunately, Uranus couldn't care less. Uranus orbited the sun as it chose, giving the best scientists of the 1800's and 1810's a great deal of mental discomfort.

That's too bad. But that's the excitement, also. Around 1820, two young mathematicians, Adams in England and Laverrier in France, predicted the existence of another planet which was supposed to perturb the orbit of Uranus. The location of this eighth planet was calculated, and indeed, after a short but careful search, it was seen.* This new planet was called Neptune. However Neptune, like Uranus, had a peculiar orbit. Then the ninth and last planet in our solar system, Pluto, was discovered.

Thus, by using perturbation theory and sticking steadfastly to Newton's laws, two new planets were discovered.

Mercury and General Relativity

For many years the behavior of the planet Mercury seemed to be very peculiar. Mercury rotates about the sun in an ellipse, but the major axis of the ellipse itself seems to rotate about 5000 seconds of arc ($\approx 1.4°$) every 100 years. By using perturbation theory, all but 43 seconds of arc could be explained. Why does the major axis of the ellipse of Mercury rotate an extra 43 seconds of arc per century? (See Fig. 5-8.) †

The search for other, undiscovered planets which might perturb Mercury's orbit proved fruitless. The unaccounted-for 43 seconds of arc had to await Einstein's theory of general relativity. This theory predicted that the major axis of the ellipses of all the planets rotate about the sun but, except for the

---

* How is it possible for an astronomer not to see a planet? If you look through a good telescope on a clear night, you'll see thousands upon thousands of objects. Most of them are stars. Others are planets, comets, meteors, asteroids, etc. It is not easy to differentiate between planets and all these other objects.

† The rotation of the major axis of an ellipse causes the whole ellipse to rotate. So, in time, a beautiful rosette pattern is traced out by the planet Mercury.

Coupling Newton's Laws of Motion and Law of Gravitation: II 73

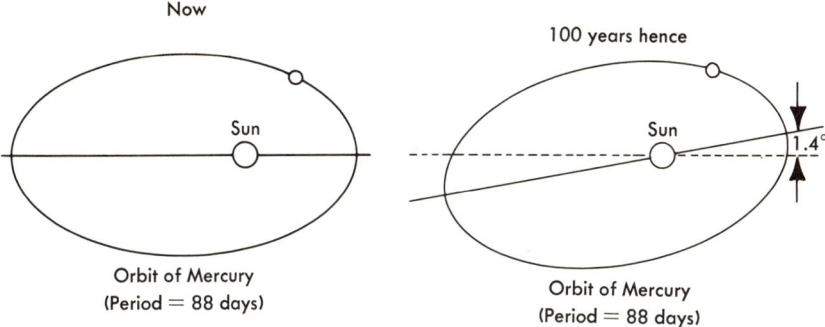

Fig. 5-8. Net rotation of the orbit of Mercury. (1 per cent of the rotation is inexplicable in terms of Newtonian gravity.)

planet Mercury, the rate of rotation is too small to be detected. Thus, Mercury's behavior provided a test for Einstein's theory.

5-5 ODDITIES OF GRAVITATION

Speed of the Gravitational Force

The sun is 92.9 million miles away from us. How long does it take for the gravitational force to travel that distance?

Newton's law of gravitation does not talk about time. Apparently, the gravitational force acts instantly. It takes the same amount of time; namely, zero seconds, for the gravitational force to travel one meter, one million meters, or one hundred million meters.

However Einstein's special theory of relativity showed that gravitational forces must travel at the speed of light, and no faster. Thus the gravitons, mentioned in Sec. 3-4, travel at the speed of light. At that speed, it takes the gravitational force from the sun 8.3 minutes to reach the earth. (See Fig. 5-9.)

Although Newton's law of gravitation may be a very good approximation to the truth, time is not included in the equation, and so Newton's law of gravitation cannot be the whole truth.

Saturation of the Gravitational Force

Consider just the earth and the sun as the only bodies in the solar system. The sun will exert a certain force on the earth. Now suppose another planet, Venus, is interposed between the earth and the sun. The sun also exerts a force on Venus.

Is the amount of force which the sun exerts on the earth diminished because of the existence of the planet Venus? In other words, does the sun have a certain amount of force it can exert and does it distribute this total

Speed of the gravitational force			
Bodies involved	Distance	Travel Time	
		Newton	Einstein
Sun-Earth	92.9×10^6 miles	0 sec	8.3 min
Earth-Moon	240,000 miles	0 sec	1.3 sec

Fig. 5-9. **Speed of the gravitational force.**

amount among all the bodies involved, or does the sun, loosely speaking, have an infinite reservoir of force? (See Fig. 5-10.)

A glance at the equation describing the gravitational force makes no mention of the existence of other bodies. The gravitational force depends only on the masses of the bodies involved and the distance between them. Thus, the force that the sun exerts on the earth is not changed one iota by the existence of Venus or by the existence of any other body in the solar system.

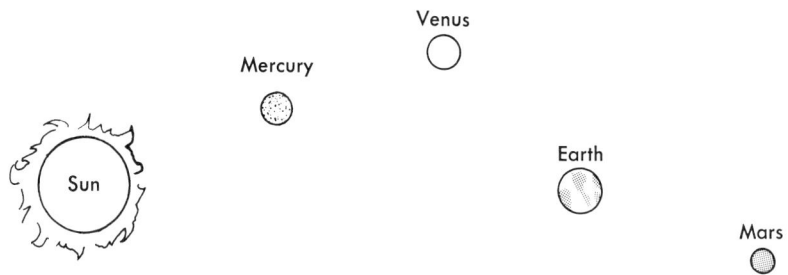

The gravitational force exerted on the earth by the sun is not reduced one iota by the presence of other planets or bodies.

Fig. 5-10. **Saturation of the gravitational force.**

This property; namely, the dependence or the independence of the force on the number of bodies involved, is called "saturation." The gravitational force is *non-saturable*. When you study the electromagnetic force,* you will find that it, too, is non-saturable. Nuclear forces, however, are saturable.

---

* See Francis E. Dart, *Electricity and Electromagnetic Fields,* Merrill Physical Science Series.

Coupling Newton's Laws of Motion and Law of Gravitation: II   75

PROBLEMS

1. Does the gravitational force between two bodies ever go to zero?
2. What effect does the principle of equivalence have on "free fall" and the sensation of weightlessness?
3. Using Eq. (3-1), show that the gravitational force on a man due to the moon changes by about 7 per cent as he goes from the side of the earth closest to the moon to the side of the earth farthest from the moon. [*Note:* let $m_{man} = 75$ kg. In one case, $d_{e\text{-}m} = (240,000 - 4000)$ miles or 3.80×10^8 m, and in the second case $d_{e\text{-}m} = (240,000 + 4000)$ miles or 3.92×10^8 m.]
4. Compute the gravitational force exerted on the earth by the moon, and compare it to the gravitational force exerted on the earth by the sun (Problem 3-4). (Let $M_{moon} = 7.4 \times 10^{22}$ kg, $M_{earth} = 6.0 \times 10^{24}$, and $d_{earth\text{-}moon} = 3.9 \times 10^8$ m.)
5. If the speed of light is 3×10^8 m/sec, how long does it take light to travel from the moon to the earth? How long does it take the gravitational force to travel from the moon to the earth?

Epilogue

A BRIEF HISTORY OF MOTION AND GRAVITATION

If it were possible for a person to live 2000 years, imagine the excitement of his life.

First the motion of all the heavenly bodies—the stars, the planets, the sun, and the moon is observed with amazing regularity. A very reasonable model of the universe (the Ptolemaic system) is introduced. According to this model, the earth is standing still in the center of the universe and all of the heavenly bodies are moving in circles about the earth. True, there are some complications due to epicyclic motion, but that is a small price to pay for being in the center of the universe.

Then, as time goes on, more and more observations are made and more and more epicycles have to be introduced to explain the observations, until the world becomes hopelessly complex. A new model of the universe (the Copernican system) is put forth, in which the earth and all the planets are moving around the sun. The earth is also spinning on its axis, and the moon is rotating about the earth. However, epicycles are still necessary in order to explain all the observations.

Then a tremendous breakthrough is made by Kepler, using data laboriously taken by Brahe (circa 1609). Kepler was able to fit the astronomical data by assuming elliptic rather than the more beautiful circular orbits. No more epicycles were necessary; rather, all of the planets, Mercury, Venus, Earth, Mars, Jupiter, and Saturn, rotate about the sun in elliptic orbits. The moon also rotates about the earth in an elliptic orbit.

About 80 years after Kepler's description of planetary motion appeared, fantastic new ideas concerning motion in general were developed. Previously, a body required a "moving force" to keep it moving. Now, a body has inertia or mass which keeps the body moving, once started. Newton finally enters the scene, and with a wonderful bit of insight and a great deal of work, he combines the problem of what is holding the solar system together with his laws of motion.

Newton, using the concept of mass, presents three laws of motion, and hypothesizes the existence of a gravitational force (a phenomenon that enables two bodies to attract each other without touching). With these two ideas, he shows that, given their initial conditions, the earth and all the other planets have no choice but to move in elliptical orbits about the sun, and that the moon has to move in an ellipse about the earth, and that all bodies fall toward the earth with the same acceleration. His laws enable man to understand a wide variety of phenomena whose interactions range from 93,000,000 miles (or 15×10^{10} m) down to 0.001 meters.

However, a glaring inadequacy in Newton's scheme is its inability to

Epilogue

solve for the exact motion of three bodies which exert forces on each other and move about simultaneously. It was true that the solution could be approximated by perturbation theory to a high degree of accuracy, but it was still an approximation. [Using Newton's laws the two-body problem can be solved exactly, and a problem involving an infinite number of bodies can be solved in a very satisfactory manner, but the solutions of problems involving 3, 4, 5 . . . bodies is approximate only.] To be sure, the gross features of a 3-, 4-, or 5-body system can be described using conservation laws.* But the problem of taking one body out of the three, and predicting its motion under the influence of the other two, is not presently solvable in a satisfactory manner. Perturbation theory is used to solve this type of problem. (See Sec. 5-4, Perturbation of the Planetary Orbits.)

Another item should be pointed out. While the Newtonian description of nature is satisfactory for a large range of problems, Newton was a man, not a god, and his laws were not of divine origin. So while they are effective, concise, and beautiful, they may not be the last word on the subject. (There are many ways of getting the same result; i.e., $4 \times 2 = 8$, $8 \times 1 = 8$, etc.)

Today, just as in the 1600's, the problem of what holds the solar system together and what the laws of motion are, are being disputed because of very slight discrepancies which are inexplicable under the Newtonian scheme; particular problems are the peculiar orbit of Mercury and the bending of light around the sun. Although these problems are discussed in more detail elsewhere,† a summary is presented here.

General relativity claims that the earth travels around the sun in an elliptical path because of the "geometry" of the space surrounding the sun. This geometry involves the four dimensions of space-time, rather than the three dimensions of ordinary space that we are familiar with. The shape of the geometry is determined by material bodies.

Consider the following: The surface of a tabletop is a flat, two-dimensional plane and the shortest line (or the straightest line, if you will) between two points on a tabletop is what we usually call a straight line. However, the surface of a sphere is a curved, two-dimensional plane, and the straightest line on a sphere is a circle. For example, the surface of the earth is spherical and so the shortest line between New York and California is a circular path over the earth's surface. Thus, for a particular position and a particular destination on the two-dimensional geometry of the earth's surface, a particular circle is the straightest line. (See Appendix 3.)

Now, according to the theory of general relativity, the sun, because of its mass, establishes a certain four-dimensional space-time geometry in its surroundings. The earth finds itself within this geometry, and for the particular position and the particular velocity of the earth on the four-dimensional space-time geometry of the sun, a particular ellipse is the straightest line. Similarly, all of the planets travel in their own particular ellipses. Using the idea of a four-dimensional space-time geometry, the peculiar orbit of the planet Mercury as well as the bending of light around the sun are explicable.

---

* See Chapter 6.
† See Isaac Maleh, *Modern Physics,* Merrill Physical Science Series.

The law of gravitation, as understood by Newton, becomes unnecessary, although the first approximation to the theory of general relativity is exactly Newton's law of gravitation. Unfortunately, the theory of general relativity is not sufficiently confirmed by experiment as yet.

So this 2000-year old person has gone from a stationary earth to a very rapidly moving earth, and finally to an earth moving in space-time.

This person has gone from the idea of divine or heavenly motion, to Newton's laws of motion and gravitational forces, and finally to the concept of motion in a four-dimensional space-time geometry with no gravitational forces.

He has gone from an earth-centered universe to an earth rotating around a medium-sized star in a medium-sized galaxy (the sun is one of 100,000 million stars in the Milky Way galaxy, and the Milky Way galaxy is just one of many millions of other galaxies).

Over this same 2000-year period, the answer to the question "What will happen if the sun disappears?" has changed markedly. (See Fig. 5-11.)

Under the Ptolemaic system (that of the earth-centered universe), it was thought that if the sun disappeared, we would have complete darkness immediately, but that the earth would remain solidly fixed in the center of the universe.

Under the Copernican system (that of the sun-centered universe), before it was learned that light travels at a finite velocity, it was thought that if the sun disappeared two things would happen: we would have complete darkness immediately, and the earth would fly off on a tangent immediately.

After we learned that light travels at a finite velocity (186,000 miles per second), it was thought that if the sun disappeared, we would see light for 8 more minutes and then we would have complete darkness, but that the earth would still fly off on a tangent immediately.

However, after the development of the theory of special relativity (1905), it was thought that if the sun disappeared, we would see light for 8.3 more minutes and then we would have complete darkness; also the earth would continue on its orbital path for 8.3 more minutes and then it would fly off on a tangent.

In a sense, each succeeding minute proves that the sun was still there eight minutes ago.

No, I guess 2000 years is not sufficient time to understand nature. We need 3000, 4000, 5000 years; then perhaps we'll know something about nature.

WHAT DO WE MEAN BY "EXPLAIN"?

What do we mean when we say Newton's laws *explain* the motion of the planets?

Briefly stated, we mean that if we accept all of Newton's laws as being true, then, within the context of Newton's laws we understand why the planets move as they do. The reason why Newton's laws are true is a separate, and, I'm afraid, unanswerable question.

If the sun should disappear, . . .

Ptolemaic system

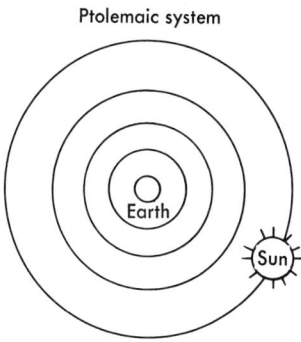

If velocity of light is *infinite*:
(1) instant darkness,
(2) earth remains fixed.

Copernican system

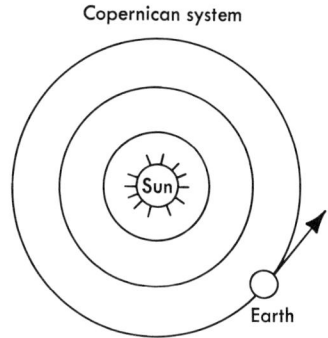

If velocity of light is *infinite*:
(1) instant darkness,
(2) earth flies off on a tangent instantly.

Copernican system

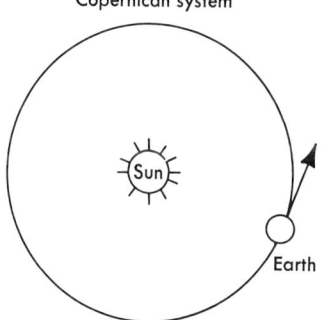

If velocity of light is *finite*:
(1) darkness in about 8 minutes,
(2) earth flies off on a tangent instantly.

Einsteinian universe

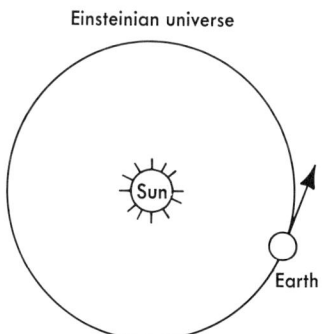

If velocity of light is *finite and* velocity of the gravitational force is *finite*:
(1) darkness in about 8 minutes,
(2) earth flies off on a tangent in about 8 minutes.

Fig. 5-11. **If the sun should disappear, . . .**

As an example of the difficulties we can get into while searching for the truth about nature, let's consider the following problem: What is the mechanism of the gravitational force?

Equation (3-1) tells us how much force a body with mass m_1 exerts on another body with mass m_2 when the distance between them is either an inch or a million miles. Even if we know how one body *could* exert a gravitational force on another body which is very far away, how does the body know how much force to exert?

Does one body compute its own mass and then the mass of the second body, then does it figure out the distance between them and finally multiply the masses together, divide by the square of the distance, and multiply by G in order to get the force it is supposed to exert on the other body? Does every body have a little computer that does this calculation to satisfy physicists? These are very difficult questions to answer, probably too difficult for human beings. Fortunately the physicist doesn't have to answer this question; he doesn't have to worry about the mechanism of the gravitational force. All he knows is that if he assumes the existence of a gravitational force as expressed in Eq. (3-1), and assumes the three laws of motion as described in Chapter 2, literally hundreds of thousands of phenomena become explicable. These ideas (gravitation and laws of motion) are adopted as axioms. Why they are true is not known.

The essential point is that although there are not very many axioms (four in all), they do explain very many phenomena. The goal is to reduce the number of axioms from four to three, from three to two, and finally (and hopefully), from two to one. With each reduction in number should come greater universality.

For example, Kepler's three laws of planetary motion describe the behavior of planets in the heavens, but they are applicable only to planets. Similarly, Galileo's experiments and conclusions concerning motion describe the behavior of bodies on earth, but they are only applicable to bodies on earth. However, Newton's laws are applicable to all bodies, those on earth and those in the heavens. Newton's laws contain both Kepler's laws and Galileo's conclusions within them. And the sum of Newton's laws (or axioms, if you will) is less than the sum of Kepler's and Galileo's.

If you study Einstein's theory of general relativity (which has less than four axioms), you will find that his theory is a partially successful attempt to reduce the number of axioms still further. However, because general relativity only concerns gravitational forces and not all kinds of forces, it is only partially successful.

Remember that the goal is to reduce the number of axioms, and simultaneously to explain more and more of nature. If and when we finally arrive at just one axiom through which all of nature can be understood, we still have the embarrassing question, "Why is that axiom true?" Is the final truth ever knowable? I think not.

Chapter 6

Conservation Laws

6-1 INTRODUCTION

Whenever one has to consider a complicated system involving many bodies moving relative to each other, colliding with each other, and interacting with each other, it is naturally very confusing. You look here and you see one thing happening, you look there and you see something else happening and then, you look again at the first region and behind your back, it has changed completely. It would be very nice to be able to say something useful about this system, but what? In the last few chapters we learned about Newton's laws. Can Newton's laws be used to predict the future of such a complicated system? No!

Unfortunately, Newton's laws are useful only if either one or two bodies are involved, or if an infinite number of bodies are involved.* Newton's laws are not useful for handling problems involving 3, 4, 5, or more bodies simultaneously.

Fortunately, even in the most complicated system some things do remain constant, and so this complicated system can be described in terms of these constant, or so-called *conserved* quantities.

6-2 AN EXAMPLE OF CONSERVATION

A Game of Poker

A very common example of something that remains the same in spite of all kinds of changes arises is a particular kind of poker game.† As you know,

---

* If an infinite number of bodies comprise a system, calculus can be used to solve the problem.

†There are many ways of playing poker. This particular game is contrived so as to illustrate some important aspects of conservation laws.

poker is a card game involving a certain number of players each of whom puts money into a pot in the center of the table and in return, is given five cards. There is a round of bidding, more money gets put into the pot, and new cards are exchanged for old cards. Then there is another round of bidding and again more money gets put into the pot. Finally, the player with the best hand wins all of the money in the pot. The game is usually complicated by various sorts of tactics; i.e., bluffing and special rules, but these need not concern us here.

Let's suppose that in our particular poker game there are six players; one player begins with $20.00, three players begin with $15.00 each, and two players begin with $10.00 each. Let's also say that no player can introduce new money into the game. Thus, when the game begins, the total amount of money in the game is $20.00 + (3 \times 15.00) + (2 \times 10.00)$ or $85.00.

In the midst of all the talk, gossip, eating, and drinking, money is transferred from the players to the pot to the other players, back to the pot, etc. For a while the game tends to oscillate: winners become losers and losers become winners. After a few hours, it becomes fairly clear which players are going to end up winners and which players are going to end up losers. However, in the midst of all this change there is something that remains constant. What is it?

It is the total amount of money in the game. The total amount of money in the game was initially $85.00 (each player had some money and there was no money in the pot). At some instant of time later, the pot may contain $5.00, five players have $10.00 each and one player has $30.00. At still another instant of time, the pot may contain $40.00, three players have $10.00 each and three players have $5.00 each. Since no new money can be introduced into the game and no money was destroyed, at any instant of time, the total amount of money in the game must be $85.00.

Important Aspects of Conservation Laws

In the above example of the constancy of the total amount of money in a poker game, three items should be mentioned explicitly.

1. A system is chosen and we concern ourselves with a particular characteristic of that system. The poker game with six players is the system, and money is the particular characteristic of the system.
2. We make sure that the system is isolated; i.e., there is no outside interference, no new money is introduced into the game, no money is burned up, etc.
3. At some instant of time, everything stops and each player, and the pot, is asked "How much money do you have now?" (The average amount of money a player has owned since the game began is of no concern; what he has now is important.)

All the contributions are added together and at any instant of time the sum must be $85.00. So, in the midst of all the changes that occur in this poker game, something remains constant: the total amount of money in the game is constant, or money is *conserved*.

Conservation Laws

Notice that we cannot predict which player will have the most money or even how much money any one player will have. All we know is that the total amount of money is $85.00. However, toward the end of the game, if there are only two players left and we know how much money is in the pot and how much money one of the players has, we can compute how much money the other player must have. (If he has more than this amount, someone is cheating.)

In physics, just as in this poker game, there are things that remain constant in the midst of change. In this chapter we will concern ourselves with just three of these things: momentum, angular momentum, and energy.

6-3 CONSERVATION OF MOMENTUM

Definition of Momentum

The momentum of any one body is defined as the mass of that body multiplied by its velocity. Mathematically,

$$(\text{Momentum}) = m \times v \qquad (6\text{-}1)$$

The momentum of a body is a measure of the linear motion of a body. Since velocity is a vector quantity, momentum is also a vector quantity.

The total momentum of a system of bodies is the sum of the momentums of all of the bodies in the system. (See Fig. 6-1.) Notice that since momentum is a vector, the sum of the momentums involves vector addition.

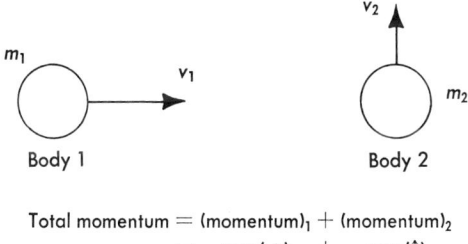

Total momentum = (momentum)$_1$ + (momentum)$_2$
          = $m_1 v_1 (\rightarrow)$ + $m_2 v_2 (\uparrow)$

Fig. 6-1. The total momentum of a system.

Statement and Analysis

The statement regarding the conservation of momentum* is,
"*If there is no net external force acting on a system of bodies, the total momentum of the system of bodies will be conserved.*"

---

* It turns out that the conservation of momentum can be proven using Newton's second law of motion. Unfortunately, lack of space forbids this demonstration.

Now, force and momentum are vector quantities, so a great deal of interpretation is necessary for the application of this conservation law. A more exact expression of the law is:

If there is no net external force in the x-direction, the total momentum in the x-direction is constant.

If there is no net external force in the y-direction, the total momentum in the y-direction is constant.

If there is no net external force in the z-direction, the total momentum in the z-direction is constant.

As before, we will only concern ourselves with two directions in this text, the x-direction and the y-direction.

In analogy with the conservation of money in the aforementioned poker game, we choose an instant of time and concern ourselves with the total momentum of all the bodies in our system at that instant of time. The only thing we have to worry about is our bookkeeping. We must separate the momentum in the x-direction from the momentum in the y-direction.

Examples of the Conservation of Momentum

(1) Let's consider an example in which we have two billiard balls, A and B, lying on what we will call the x-axis. (See Fig. 6-2.) B is at rest, and A

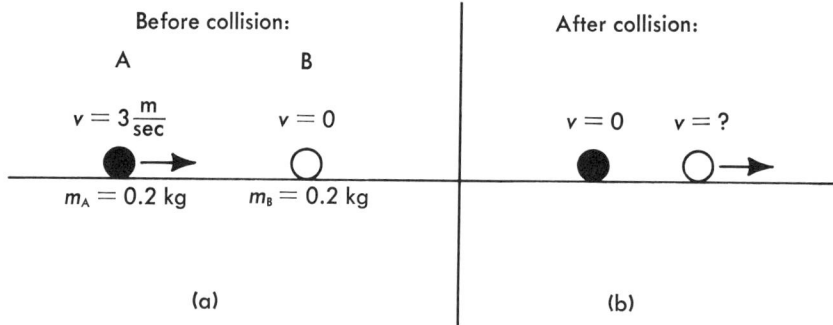

Fig. 6-2. An example of the conservation of momentum.

is moving toward B with an x velocity of 3 m/sec. Both billiard balls have a mass of 0.2 kg. The balls approach each other, collide, and separate. If we consider both balls as our system, there is no net external force acting on our system.* Thus, in spite of the collision, the total momentum of our system is constant.

---

* Both surface friction and air friction are assumed to be zero.

Conservation Laws

Before the collision:

x-direction	y-direction

(Total momentum) = (momentum)$_A$ + (momentum)$_B$
(Momentum)$_A$ = $m_A v_A$
                 = 0.2 × 3.0
                 = 0.6
(Momentum)$_B$ = $m_B v_B$
                 = 0.2 × 0.0 = 0.0
(Total momentum) = 0.6 + 0.0
                         = 0.6
	Nothing happens in the y-direction.

No matter what happens after the collision, the total momentum in the x-direction must be 0.6 units of momentum.* (Incidentally, the total momentum in the y-direction is zero initially, and since there is also no force in the y-direction, the total momentum in the y-direction must also remain zero.) Let's suppose that ball A stops immediately after the collision. What is the velocity of ball B after the collision?

After the collision:

x-direction	y-direction

(Total momentum) $_{final}$ = (total momentum) $_{initial}$
(momentum)$_A$ + (momentum)$_B$ = 0.6
          0.2 × 0 + 0.2 × v = 0.6
                     0.2 × v = 0.6
                           $v = 3.0 \frac{m}{sec}$
	Nothing happens in the y-direction.

Thus, ball B moves with a velocity of 3.0 m/sec in the x-direction.

(2) Let's consider another example involving a bullet and a block of wood. We will label the bullet body 1, and the block of wood body 2. The data involved are in Fig. 6-3 (a) and (b).

---

* The unit of momentum in the MKS system is $\frac{kg \times m}{sec}$.

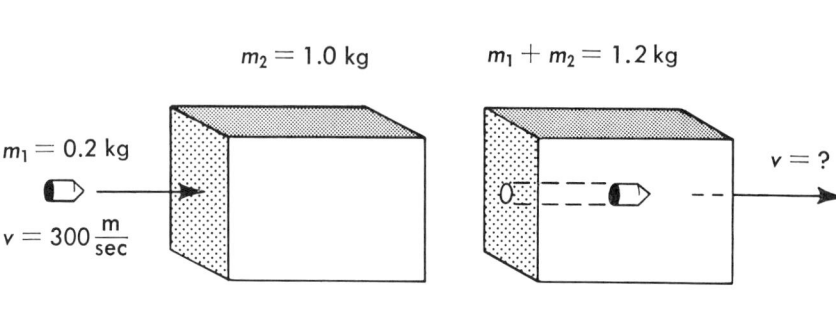

Fig. 6-3. An example of the conservation of momentum.

Before the collision:

x-direction	y-direction
(Total momentum) = (momentum)$_1$ + (momentum)$_2$ (Momentum)$_1$ = $m_1 v_1$ = 0.2 × 300 = 60.0 (Momentum)$_2$ = $m_2 v_2$ = 1.0 × 0.0 = 0.0 (Total momentum) = 60.0 + 0.0 = 60.0	Nothing happens in the y-direction.

Thus, no matter what happens after the collision, the total momentum in the x-direction must be 60.0 units of momentum. Since the bullet remains in the block after the collision, our system has become one body whose mass is the sum of the mass of bullet and the mass of the block of wood; i.e., 1.2 kg.

After the collision:

x-direction	y-direction
(Total momentum)$_{\text{finally}}$ = (total momentum)$_{\text{initially}}$ $(m_1 + m_2)v$ = 60.0 1.2 v = 60.0 v = 50.0 $\dfrac{\text{m}}{\text{sec}}$	Nothing happens in the y-direction.

Conservation Laws

Notice that in spite of the noise produced, in spite of the hole made in the block, and in spite of the smoke created,* the total momentum in the x-direction remained constant.

Before going further, I want to point out a very subtle but important aspect of these examples in particular, and of conservation laws in general. I claimed that nothing happened in the y-direction in both examples. That was a guess. The conservation of momentum law does not quite say that nothing happens in the y-direction. Rather, the conservation of momentum law says that whatever happens in the y-direction, the total momentum in that direction must be zero. For example, the conservation of momentum law is not violated if, in the first example, marble A moves in the y-direction. The conservation of momentum law demands, however, that if marble A does move in the y-direction, marble B must also move in the y-direction so that the total momentum in the y-direction remains zero.

If the conservation of momentum law allows more than one thing to occur, how then does nature choose only one final activity? The answer lies in the fact that there are many different conservation laws. Each conservation law allows a class of final states, and nature chooses that single state which is compatible with all the conservation laws. (See Fig. 6-4.)

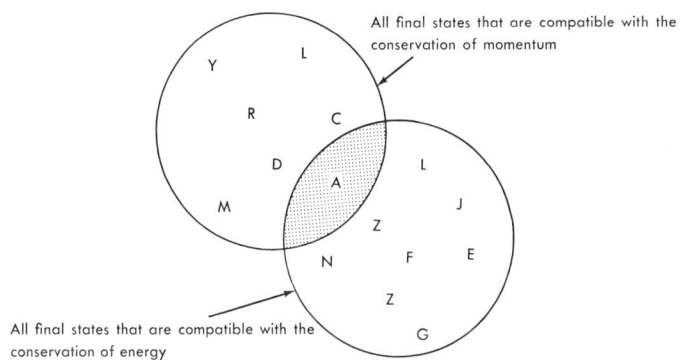

The final state chosen by nature is state A.

Fig. 6-4. **The superposition of conservation laws determines the final state of the system.**

Use of Conservation of Momentum

To see the usefulness of the conservation laws, let us return to the problem of a man in outer space. As mentioned earlier (Sec. 5-2), he wants to change his velocity but there is no object on which he can exert a force so that the reaction force can change his velocity. His knowledge of the principle of the conservation of momentum saves the day.

---

* The smoke created means that some atoms were lost in our bullet-block system, but fortunately those atoms do not have enough mass to change the answer significantly.

Suppose for the moment that he is standing still, far away from any body. If he wants to go toward his left, he can reason as follows: My total initial momentum in the right-left direction (or x-direction) is zero, and there are no net forces acting on me in the x-direction; therefore, my total momentum in the x-direction must always be zero. I can take off my coat and throw it to my right giving it a momentum $m_{coat} \times v_{coat}$ in the negative x-direction.* Since my total x-momentum must be zero, I must go toward my left so that

$$0 = m_{coat} v_{coat} + m_{me} v_{me}$$

In particular, if $m_{coat} = 2$ kg, $v_{coat} = -3$ m/sec, and $m_{me} = 60$ kg, then

$$0 = 2 \times (-3) + 60 v_{me}$$
$$0 = -6 + 60 v_{me}$$
$$60 v_{me} = 6$$

or

$$v_{me} = 0.1 \frac{m}{sec}$$

The velocity is small, but it is larger than before. (See Fig. 6-5.)

In this manner, our spaceman can go in any direction he likes.

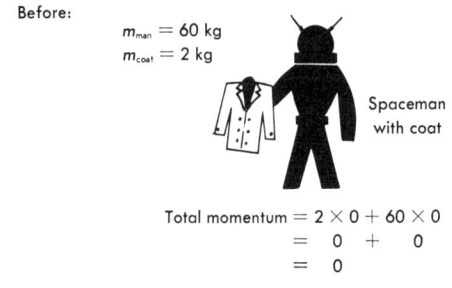

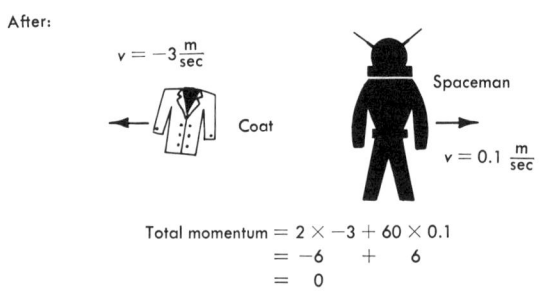

Fig. 6-5. Using the conservation of momentum in outer space.

A more practical situation is one in which the man is orbiting the earth. By using tanks of compressed gas and regulating the direction, amount, and velocity of the emitted gas, he can go wherever he wants to go.

---

* The positive x-direction is toward the spaceman's left.

6-4 CONSERVATION OF ANGULAR MOMENTUM

Definition of Angular Momentum

Just as *momentum* is a measure of the linear motion of a body, *angular momentum* can be considered as a measure of the circular motion of a body.

For an object moving in a perfect circle, the angular momentum equals the radius of the circle times the momentum of the object. Mathematically,

$$\text{(Angular momentum)} = r \times mv$$
$$= mvr \tag{6-2}$$

Angular momentum is a vector, just like momentum, but although the magnitude is clearly defined (mvr), the direction is not obvious. By definition, the direction associated with angular momentum is determined by the thumb of the right hand. (See Fig. 6-6.)

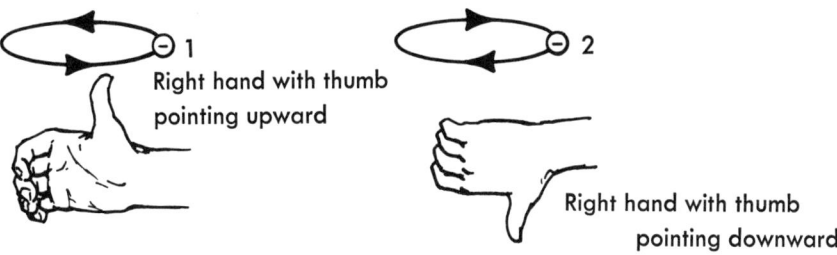

Rotating particles. (Direction of angular momentum is described by curling fingers of right hand and sticking thumb out.)

Fig. 6-6. The direction of angular momentum.

For an object moving in an elliptic path, the angular momentum equals the perpendicular radius multiplied by the momentum of the object. (See Fig. 6-7.) Mathematically,

$$\text{(Angular momentum)} = r_\perp \times mv$$
$$= mvr_\perp \tag{6-3}$$

Statement and Analysis

The conservation law concerning angular momentum is,

"*If there is no twisting force acting on the system, the total angular momentum of the system is constant.*"

A common example of the conservation of angular momentum occurs

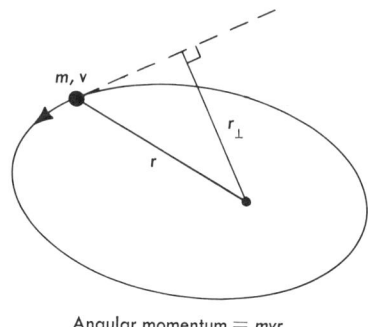

Angular momentum = $mvr_\perp$

Fig. 6-7. Angular momentum in an elliptical orbit.

in an ice-skating rink. (See Fig. 6-8.) The skater begins turning so that each part of her is rotating about in a circle. Therefore, each part of our skater has a certain *mvr* and the sum of these *mvr*'s represents her total angular momentum. Since there is no twisting force being exerted on our skater, her total angular momentum is constant. Now, if our skater pulls in her arms, the radius associated with some of her rotating parts has decreased, so that the velocity must increase in order that the sum of the *mvr*'s remain the same.

Total angular momentum is the sum of the *mvr*'s.

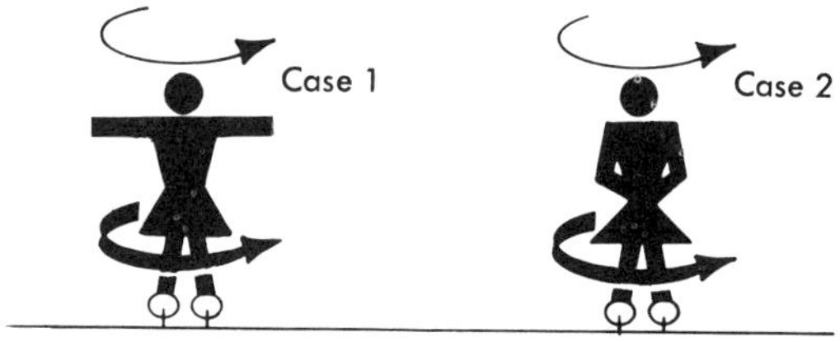

Case 1   Case 2

The *r*'s in case 2 are less than the *r*'s in case 1; therefore, the *v*'s in case 2 must be greater than the *v*'s in case 1 so that the sum of the *mvr*'s remains the same.

Fig. 6-8. An example of the conservation of angular momentum.

6-5 CONSERVATION OF ENERGY

Definition and Remarks on Energy

A scoop of ice cream has about 200 Calories and a dash of whipped cream has about a million Calories. What exactly is a Calorie? Just as an inch is a unit of distance, a Calorie is a unit of energy, and somehow too many Calories; i.e., too much energy in the form of food, causes us to weigh too much. Although energy is almost an everyday term, it actually represents a very abstract and complex idea.

Conservation Laws   91

One problem is the fact that energy appears in so many different forms:

Potential energy	Energy due to position.
Kinetic energy	Energy due to motion.
Heat energy	Energy due to atomic or molecular motion.
Sound energy	Energy in sound waves.
Chemical energy	Energy in chemical bonds. (The energy in foodstuffs is really energy in the form of chemical energy.)

Electrical energy, etc.

Statement and Analysis

In this text, for simplicity, we will limit ourselves to systems that have only potential energy (P.E.) and/or kinetic energy (K.E.). With this limitation, the law of the conservation of energy is,

"The net work done on a system equals the change in energy of the system."

Now the phrase "work done on a system" needs to be defined. If the system is a body, then the work done on the body is defined as the net force acting on the body multiplied by the distance the body moves in the direction of this net force. If the force is at right angles to the motion of the body, no work is done on the body. Also, if the body does not move, no work is done on the body regardless of the size of the force. (See Fig. 6-9.)

There is no work done on the object.

A man is pushing with a force of 100 lbs on a building. If the building does not move, the man has done no work <u>on</u> the building.

There is work done on the object.

A man is pulling on a block. If the block moves in the direction of the force, the man is doing work <u>on</u> the block.

Fig. 6-9. **The concept of work.**

We see that the definition of "work" is not at all obvious. A man can push against a wall with a force of 100 lb for 8 hours, but because the wall will not move, he has done no work on the wall. The reason he is sweating and tired is because his muscles were straining to apply the 100-lb force. The food he ate supplied the energy so that his muscles could apply the force, but the wall did not move in the direction of the force, so he did no work on the wall, and the energy of the wall did not change.

Potential Energy

Let's consider a suitcase weighing 30 lb (135.0 nt) sitting on the surface of the earth. We may say that the earth is exerting a gravitational force of 135.0 nt on the suitcase and that our system is the earth-suitcase combination. A man comes along, and for some unknown reason, lifts the suitcase one meter off the ground. The man represents an external force which has done work on our system. The man has done no work on the earth because it did not move, but the man has done work on the suitcase. (See Fig. 6-10.)

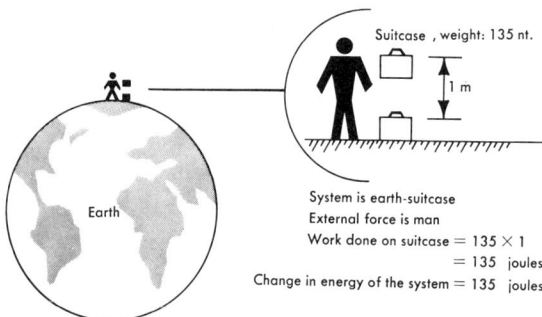

Fig. 6-10. Man changing the energy of our earth-suitcase system.

In MKS units, the work done on the suitcase is 135.0 newtons times 1 meter, or

$$W = 135.0 \text{ nt} \times 1 \text{ m}$$
$$= 135.0 \text{ joules}$$

Thus, the change in energy of our earth-suitcase system is 135.0 joules. (Notice the introduction of a new word, *joule*. A joule, abbreviated j, is a unit of energy, and one joule is defined as the work done by one newton moving an object through a distance of one meter; i.e., 1 joule = 1 nt × 1 m. It turns out that 1 Calorie = 4200 joules.)

Now suppose that the man carries the suitcase three blocks. He has done no new work on our earth-suitcase system. Why not? The man, in walking these three blocks, swings the suitcase to and fro. Relative to the ground the suitcase is always moving forward, but the man is either pushing it forward or pulling it back. When he pushes it forward he is doing positive work on the suitcase; when he pulls it back he is doing negative work on the suitcase (the force is opposite the direction of travel). The negative work negates the positive work and so, at the end of three blocks, no work is done on the

Conservation Laws

suitcase.* It is true that the man has to hold the suitcase up; i.e., exert a vertical force to keep the suitcase from falling toward the earth, but this vertical force does no work on the suitcase because the suitcase does not move in the vertical direction. Of course, the man is perspiring, but that is the result of the chemical energy that is being utilized to keep his muscles stressed. The perspiration is an indication of work being done on the man's muscles, not of work being done on the suitcase.

The energy in our earth-suitcase system is dependent only on the distance between the earth and the suitcase. Suppose that the man drops the suitcase before he walks three blocks horizontally. The suitcase hits the ground with a thud; some noise, some bending, and some crushing is produced. If he dropped the suitcase after he walked the three blocks horizontally, the same amount of noise, bending, and crushing would occur. If there were more energy in the system the second time, the resultant collision would have been more disastrous. Therefore, it is not unreasonable to conclude that no work is done on the suitcase during the horizontal motion.

The sort of energy we have been talking about is called *potential energy* (P.E.), since it has to do with the relative position (not the velocity) of objects in a particular system.

In general,† the potential energy of a body a distance h above the ground can be considered to be its weight (mg) multiplied by its height (h); or

$$\text{P.E.} = mgh \text{ joules} \qquad (6\text{-}4)$$

Notice that, in a subtle manner, the zero level of P.E. has been chosen; i.e., if $h = 0$, P.E. $= 0$. This means that the potential energy on the ground is zero. Below the ground, the P.E. would be negative, and above the ground the P.E. would be positive. However, keep in mind the fact that the level from which h is measured is arbitrary. (See Fig. 6-11.) This level is chosen according to the convenience of the problem. It is the *change* in potential energy that is important, and so it is the change in h that is important—not the zero level of h.

Kinetic Energy

The other form of energy that is useful for our purposes is energy due to motion, or *kinetic energy* (K.E.). (See Fig. 6-12.) In this case, when a man does work on a body, the work expresses itself in the form of motion. The K.E. of a body with mass m, and velocity v, is

$$\text{K.E.} = \tfrac{1}{2} mv^2 \text{ joules}‡ \qquad (6\text{-}5)$$

---

* In theory, a very short impulse is necessary to start the suitcase moving horizontally and it will continue to move horizontally forever (neglecting air friction). After three blocks, another impulse stops the suitcase and so again, no work is done on the suitcase.

† There are many forms of P.E. (for example, a compressed spring), but we will concern ourselves only with the form presented here.

‡ Although this formula can be derived by combining Newton's second law of motion, the distance, velocity, acceleration formulas in Sec. 1-2, and the definition of work, space limitation forbids the demonstration.

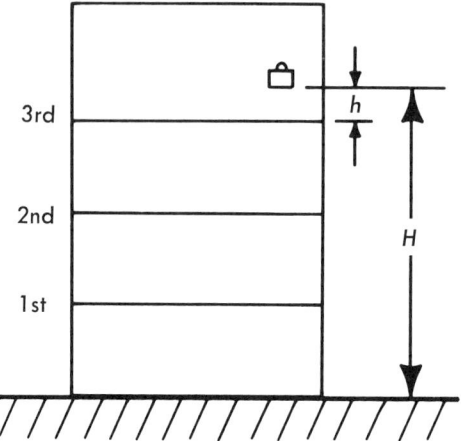

P.E. = mgh (if viewed from the 3rd floor)

P.E. = mgH (if viewed from the ground floor)

Fig. 6-11. The zero level of potential energy is a matter of convenience.

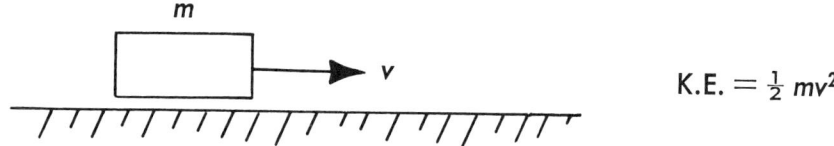

K.E. = $\frac{1}{2} mv^2$

Fig. 6-12. Energy due to motion (kinetic energy).

Example of Conservation of Energy

A Simple Pendulum:

We are now ready to consider a problem. A pendulum of mass 1 kg is raised 1 m from its lowest point and is then released. The pendulum picks up speed, attaining a maximum velocity at its lowest point, and then slows down as it begins its climb up to 1 m on the other side of the center.

The system we are talking about is the pendulum and the earth. A man, external to this system, performed work on the system by pushing the pendulum and the earth one meter farther apart than before. After this initial work, the man disappears. The gravitational force of the earth on the pendulum, as well as the force in the string holding the pendulum bob, causes the pendulum to swing back and forth. At all points in its travel, the pendulum has a certain amount of P.E. and a certain amount of K.E. (See Fig. 6-13.)

Conservation Laws

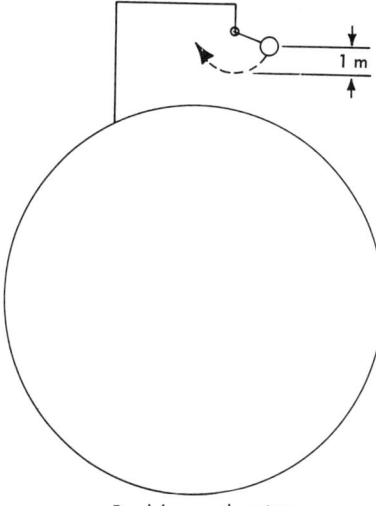

Pendulum-earth system

Fig. 6-13. A simple pendulum.

What can be said about the energy of this pendulum-earth system? Is the P.E. constant? No! Is the K.E. constant? No! If no work is done on the system (after the initial raising of the pendulum bob), the conservation of energy law says that there is no change in the total energy of the system. Thus, the total energy of our pendulum-earth system is constant.

There are only two forms of energy to be considered here—potential energy and kinetic energy. If we can calculate the P.E. and the K.E. of our system at any one point, their sum should be the same for all points.

At point A (when $h = 1$, see Fig. 6-14):

Potential energy	Kinetic energy
If the lowest level is considered the $h = 0$ level, then the P.E. at point A is $$\text{P.E.} = mgh$$ $$= 1 \times 10.0 \times 1$$ $$= 10.0 \text{ j}$$	Since the pendulum was released at point A with no initial velocity, the K.E. at point A is $$\text{K.E.} = \tfrac{1}{2}mv^2$$ $$= \tfrac{1}{2} \times 1 \times 0^2$$ $$= 0.0 \text{ j}$$

Thus,
$$\text{Total energy} = (\text{T.E.})_{h=1} = \text{P.E.} + \text{K.E.} \quad \text{(6-6)}$$
$$= 10.0 + 0.0$$
$$(\text{T.E.})_{h=1} = 10.0 \text{ j}$$

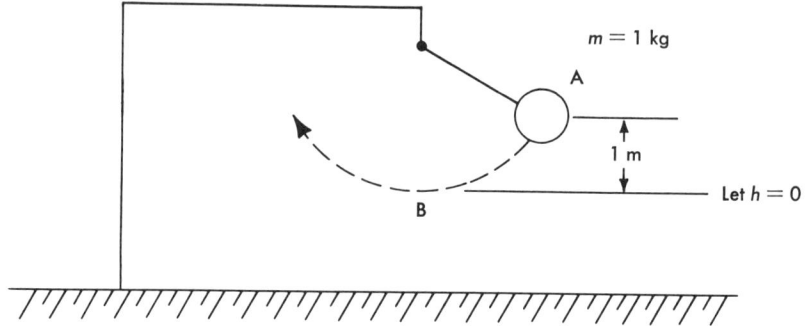

If the h = 0 level is as shown, at point A, the P.E. is 10.0 j and the K.E. is 0.0 j. At point B, the P.E. = 0.0 j and the K.E. is 10.0 j.

Fig. 6-14. An example of the conservation of energy.

Since the total energy at all points and for all time* must be 10.0 joules, we can use this information to find the velocity of the pendulum at the bottom of the swing.

At point B (when $h = 0$):

Potential energy	Kinetic energy
P.E. = mgh $= 1 \times 10.0 \times 0$ $= 0.0$ j	K.E. = $\frac{1}{2} mv^2$ $= \frac{1}{2} \times 1 \times v^2$ $= 0.5 v^2$ j

Now,

$$(T.E.)_{h=0} = (T.E.)_{h=1}$$
$$(P.E. + K.E.)_{h=0} = 10.0$$
$$0.5 v^2 = 10.0$$
$$v^2 = 20.0$$
$$v = 4.5 \frac{m}{sec}$$

In the above example, although the total energy was conserved, the composition of the energy was changing constantly. It went from pure potential energy (when $h = 1$ m) to pure kinetic energy (when $h = 0$ m), back to pure P.E., etc., ad infinitum. In between these two extremes, the total energy was shared between potential energy and kinetic energy. The pendulum is constantly moving, one side to the other, and is changing its velocity and direction; however, the total energy remains the same, and, because of this, many features of its motion can be understood.

---

* We consider an ideal case, with no air, no friction, etc.

Conservation Laws

If the pendulum were 2 m above a table top, and the table top was taken to be $h = 0$, would the results be the same? Yes! The only change is the fact that the P.E. would not be zero at the bottom of the swing.

The student should try this problem. All that has to be done is to set the total energy at the top of the swing equal to the total energy at the bottom of the swing. [*Hint:* The total energy at the top of the swing is no longer 10.0 j.]

An interesting question is, "Since the pendulum-earth is our system and no external forces are acting on our system, the total momentum should be constant. However, the momentum of the pendulum varies as it swings to and fro. What happened to the conservation of momentum law? (See Fig. 6-15.) Just as the pendulum is swinging to and fro, the earth is also swinging to and fro. The total momentum of our pendulum-earth system remains constant. (Recall that the mass of the earth is about a million-billion-billion kilograms, so it doesn't need a very large velocity to cancel the momentum of the pendulum.)

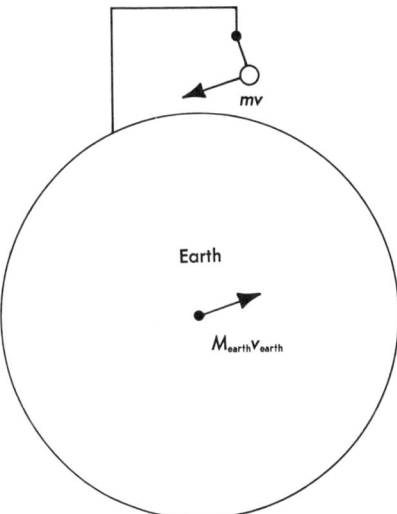

The earth and the pendulum are "rocking" back and forth so that the total momentum of the earth-pendulum system is constant.

Fig. 6-15. The simple pendulum and the conservation of momentum.

6-6 HEAT AS A FORM OF ENERGY

In a real-life situation, there is air, there is friction, etc., so that the pendulum will eventually stop and remain at its lowest point. The total energy at this point would be zero. The question is, "What happened to the 10.0 joules of energy which the pendulum had initially?"

It went into heating the air and the bearings to which the pendulum cord was attached. Each movement of the pendulum brushed the air aside, so the pendulum did work on the air. The bearings also suffered; the constant rubbing heated them and reduced their effective lifetime. So the 10.0 joules of energy were not lost, but were converted into heat. This is the ultimate fate of all forms of mechanical energy.

Consider the problem of raising a 2-kg book 5 meters off a table top. (See Fig. 6-16.) It has gained potential energy. If you let it go, it will pick up speed and finally hit the table with a bang and stop moving. Some of its potential energy has been converted into sound energy (the bang), some into heat energy (increasing the temperature of the table top as well as the book), and some has been used to crush the book.

$$\text{Total initial energy} = mgh$$
$$= 2 \times 10.0 \times 5$$
$$= 100.0 \text{ j}$$

$$\text{Total initial energy} = \begin{cases} \text{Total energy just before} \\ \text{hitting the table} \end{cases}$$

$$100.0 = (\text{P.E.})_{h=0-} + (\text{K.E.})_{h=0-}$$
$$= 2 \times 10.0 \times 0 + \tfrac{1}{2} \times 2 \times v^2$$
$$100.0 = 0 + v^2$$
$$v = 10.0 \frac{\text{m}}{\text{sec}}$$

[*Note:* The subscript $h = 0-$ means that h is almost, but not quite, zero.]

Therefore, the velocity of the book just before it hit the table was 10.0 m/sec. After hitting the table all the energy (100.0 joules) is converted

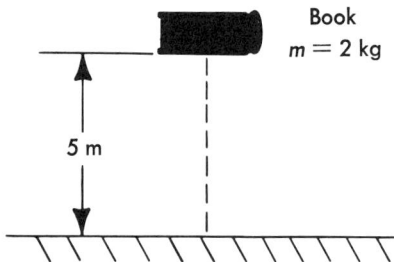

Before dropping:
Total energy of book-earth system = 2 × 10 × 5 = 100.0 j

After dropping:
Total energy of book-earth system = 0.0 j, and 100.0 j of energy has been converted into sound energy, heat energy, and energy used to deform the book (crushing).

Fig. 6-16. Heat as a form of energy.

Conservation Laws

into sound energy, into heat energy and/or is used to deform the book. There is also a very small amount of energy given to the air as the book falls and brushes the air aside.

6-7 REVIEW

Conservation laws are a tool for handling complex systems. They provide an overview of what might happen to a system. Also, of the infinity of possible final states a system may enter, each conservation law allows a certain set. Nature choses that one final state which satisfies all of the conservation laws simultaneously. For example, the conservation of momentum alone may allow a whole class of final states; however, from this class of final states, the one that also satisfies the conservation of energy will be the one chosen by nature.

We can view the Newtonian approach to physics as a step-by-step prediction of how bodies go from one state to another; i.e., forces are exerted, bodies accelerate, distance is traversed, etc. However, the conservation-law approach does not concern itself with how bodies go from one state to another; the conservation-law approach dictates certain characteristics of the final state in lieu of certain characteristics of the initial state.

If we have an isolated system, by definition, no work can be done on it, so its total energy must remain constant. Since the whole universe can be considered as an isolated system (i.e., there is nothing outside the whole universe), the total energy of the universe is constant. The composition of this energy; e.g., P.E., K.E., heat energy, etc., can change, but the total energy remains constant.

Of all the conservation laws we have discussed, by far the simplest one to use is the conservation of momentum. The momentum of a body is clearly defined as mv, and since momentum is a vector quantity, the only difficulty is in the bookkeeping; i.e., we must separate the x-, y- and z-components of momentum. The second simplest conservation law is the conservation of angular momentum. Angular momentum is also clearly defined, and again the difficulty is in bookkeeping since it is a vector quantity. The conservation of energy is difficult to use in a real-life problem because energy takes on so many different forms. Although some simple problems can be treated very well, real-life problems are not simple.

There are many other conservation laws that are sometimes simpler and sometimes more complex than the laws of conservation of momentum, angular momentum, and energy; e.g., conservation of electric charge, conservation of parity, etc. Perhaps you will study them later in your course.

PROBLEMS

1. What is the momentum of a 3000-kg truck moving with a velocity of 1 m/sec? What is the momentum of a 500-kg car moving with a velocity of 6 m/sec? What is the momentum of a 1-kg ball moving with a velocity of 300 m/sec?
2. A block of wood is at rest, and a 0.5-kg bullet is moving toward the block at

200 m/sec in the x-direction. The bullet goes right through the block, exiting with a velocity of 100 m/sec in the x-direction. What is the resulting velocity of the block? (Let $m_{block} = 5$ kg, and assume that there is no friction between the block and the ground.)

3. What is the angular momentum of the earth in its orbit about the sun? (Let $M_{earth} = 6.0 \times 10^{24}$ kg, $d_{sun\text{-}earth} = 1.5 \times 10^{11}$ m, and $v = 3.0 \times 10^4$ m/sec. Assume a purely circular orbit.)

4. What is the kinetic energy of the earth in its orbit about the sun? (Use the data in Problem 3 above.)

5. A compressed spring has what sort of energy contained within it, P.E. or K.E.?

6. A boy, mass 30 kg, is on a swing. He is raised 1.25 m above his lowest point. Using the conservation-of-energy principle, what velocity will he attain at his lowest point? Repeat the problem for a hypothetical 3000-kg man. Why is the answer identical for a 30-kg boy or a 3000-kg man?

7. Do conservation laws tell you exactly what will happen to a system of bodies?

Chapter 7

Thermodynamics

7-1 INTRODUCTION

Thermodynamics, commonly called heat, is the study of matter and energy and their interrelationship. It is a very broad subject so only selected topics will be discussed. This chapter will cover the composition of matter, the atomistic nature of matter, the states of matter, and temperature. In the following chapter we will concentrate on one particular state of matter—the gaseous state. In that chapter, we will discuss the real properties of a gas, and we will then present the predicted properties of a gas as determined from the so-called kinetic theory of gases. Finally, we will compare the real properties of a gas with its predicted properties.

7-2 THE COMPOSITION OF MATTER

There are air, water, and sand. There are metal products, wooden products, and paper products. There are diamonds, plastics, and glassware. In just about one minute it is possible to name a hundred different kinds of matter all around us. Are all these things really different, or are they just different states of the same thing?

Well, an old Greek concept of matter, which turned out to be true, is that there are a certain number of basic *elements,* and that all matter is composed of various amounts of each of these basic elements. At one time the basic elements were thought to be Earth, Water, Air, and Fire. Wood, for example, was thought to be composed of lots of earth, some water, a little air, and very little fire. If the wood were reddish, it contained more fire; if it were lighter, it contained more air, etc. (See Fig. 1-2 in Chapter 1.) Any substance which was composed of various amounts of these elements was

HEAT
(Thermodynamics)

Matter: Types	Microscopic Characteristics (See Secs. 7-2 and 7-3.)
Elements	Same kinds of molecules Same kinds of atoms within each molecule
Compounds	Same kinds of molecules Different kinds of atoms within each molecule
Mixtures	Different kinds of molecules

States of Matter:
- Solid
- Liquid
- Gas

Heat and Temperature:

Heat	Temperature
A form of energy	A measure of the potential energy flow between two bodies. (Also, in the case of a gas, a measure of the average K.E. of the molecules.)

The Gaseous State:

Real gases	Kinetic theory of gases
Experiments	Model of a gas
↓	↓
P, V, T Measurements	Mathematics (Temperature and K.E.)
↓	↓
Facts ←	→ Predictions

AGREEMENT
(almost)

Thermodynamics

called either a *compound* or a *mixture*. Although earth, water, air, and fire lost their place as elements, the concept of elements, compounds, and mixtures is true and has survived.

Elements, Compounds, Mixtures

The current scheme concerning matter utilizes five words and ideas: homogeneity, substances, elements, compounds, and mixtures.

If every portion of a particular piece of matter is the same as every other portion, then this piece of matter is *homogeneous*. For example, in New York City in the 1930's and 1940's, milk used to be delivered in glass bottles and the cream was always at the top of the bottle. (Cream is lighter than non-fat milk.) We always had to turn the bottle upside down in order to "homogenize" the milk. Prior to the mixing, the milk was inhomogeneous because a portion near the top would contain almost all cream and no non-fat milk, while a portion near the bottom would contain all non-fat milk and no cream. After the mixing, however, the milk was homogeneous in that every portion would contain the same amount of cream and non-fat milk. Today, of course, we can buy homogenized milk directly. It is difficult enough to discuss matter which is homogeneous, so we will completely disregard matter which is inhomogeneous.

There are many different kinds of homogeneous matter, and we will call each kind of homogeneous matter a *substance*. Homogenized milk is a substance, orange juice is a substance, and uniformly polluted air is a substance. Each substance has its own particular properties. For example, the substance water has a melting point of 0°C (melting ice,) a boiling point of 100°C (boiling water), and a certain density. Gold has a melting point of 1063°C, a boiling point of 2600°C, and a certain density. Generally, the melting point, boiling point, and density of one substance is different from the melting point, boiling point, and density of any other substance. (See Fig. 7-1.) Now, substances can be either *elements, compounds,* or *mixtures*.

How can we tell whether a particular substance is an element, a compound, or a mixture?

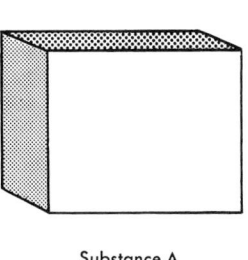
Substance A

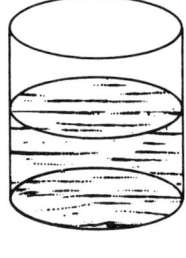
Substance B

Fig. 7-1. Every substance has its own melting point, boiling point, density, etc.

The basic difference between elements, compounds, and mixtures lies in the differences in their respective "building blocks." To understand this more fully, let's consider the building blocks of various types of substances.

7-3 THE ATOMISTIC NATURE OF MATTER

Atoms and Molecules

At present, there are 103 different elements known, and these elements combine in various ways to form the substances of this world.

What is an element like? Or, what are the building blocks of an element?

All of the elements are made of atoms, and different elements have different kinds of atoms associated with them. That is to say, one pound of the element gold is composed of millions upon millions of gold atoms and one pound of the element hydrogen is composed of millions upon millions of hydrogen atoms. But the gold atom is different from the hydrogen atom. Also, there are oxygen atoms, silver atoms, carbon atoms, etc.; each of the 103 different elements has its own particular atom. Since matter is composed of different kinds of elements, and different kinds of elements are composed of different kinds of atoms, matter is really composed of collections of different kinds of atoms. (A definition: When the building block of a substance is composed of two or more atoms, either the same atom or different atoms, the building block is called a molecule.) Thus, the building blocks of matter are either atoms or molecules. (See Fig. 7-2.)

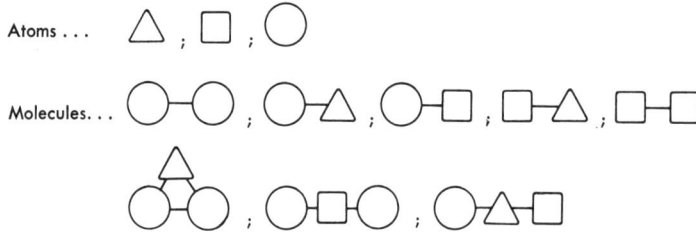

Notice that some molecules are composed of the same kinds of atoms while other molecules are composed of different kinds of atoms.

Fig. 7-2. **Atoms and molecules.**

With this concept of molecules, it is easy to define the difference between elements, compounds, and mixtures.

If all of the molecules of a substance are identical, then the substance is either an element or a compound. (It is an element if the molecule is composed of the same type of atom; it is a compound if the molecule is composed of different types of atoms.) However, if all the molecules of a substance are not identical, then the substance is a mixture. (See Fig. 7-3.)

Hydrogen, for example, is composed of molecules, but because the hydrogen molecule is composed of two hydrogen atoms, hydrogen is an element. Water is also composed of molecules, but the water molecule is

Thermodynamics

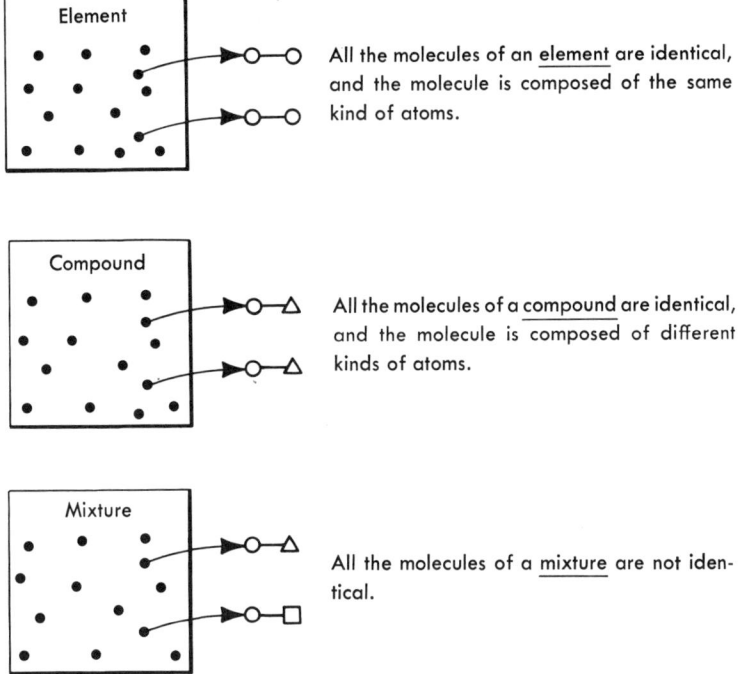

Fig. 7-3. **Elements, compounds, and mixtures.**

composed of two hydrogen atoms and one oxygen atom, so water is a compound. Sugar water is also composed of molecules but there are two different kinds of molecules in sugar water. One kind is a water molecule and another kind is a sugar molecule. Sugar water, then, is a mixture, because it is composed of different kinds of molecules.

Around 1800, one way of defining the difference between elements, compounds, and mixtures depended solely on the possibility or the manner of decomposition. In 1781 A. Lavoisier, a French chemist, presented an empirical and therefore utilitarian definition of an element, as well as a list of the known elements of his time. Lavoisier said, "A substance is an element if it cannot be decomposed into other substances." If a substance can be decomposed, it is either a compound or a mixture. Aside from guessing, how can we tell whether it is a compound or a mixture? Because a mixture contains different kinds of molecules, it can be decomposed fairly easily. A compound, however, contains the same kind of molecule, and so the only way to decompose a compound is to tear the molecule apart; i.e., a chemical reaction is necessary. So, if the substance could be decomposed by means of a wire screen or filter paper or by other so-called mechanical means, then the substance was a mixture. If the substance had to be decomposed electrically or chemically, then a chemical reaction was involved and the substance was a compound.

Hydrogen and oxygen, for example, are elements because they cannot be decomposed mechanically, electrically, or chemically into any other substance. Gold, silver, lead, copper, and iron are also elements, because they cannot be decomposed into any more basic substances. However, sugar water can be decomposed mechanically; e.g., with filter paper, into sugar and water. Thus, sugar water is a mixture. Water itself is a compound, because although it cannot be decomposed mechanically, it can be decomposed electrically. Electric current will decompose water into hydrogen and oxygen.

Although the "decomposition" approach to defining elements, compounds, and mixtures is essentially *ad hoc,* it is very useful in the laboratory. The more basic difference between elements, compounds, and mixtures lies, however, in the differences in their respective building blocks.

Proof of the Atomistic Nature of Matter

Elements are composed of atoms. But no one has ever seen an atom, so how do we know that elements are composed of atoms? Matter is composed of molecules. But no one has ever seen a molecule, so how do we know that matter is composed of molecules?

John Dalton of Manchester, England

The first hint of the atomistic nature of the elements, and, therefore, the atomistic nature of matter, was detected by John Dalton (1776-1844) of Manchester, England. He noted that when elements combined to form compounds, or when compounds were decomposed into their elements, fixed ratios appeared. For example, if somehow we can decompose 28 pounds of carbon monoxide, we will get 12 pounds of carbon and 16 pounds of oxygen. But if we decompose 44 pounds of carbon dioxide, although we still get only 12 pounds of carbon, we get 32 pounds of oxygen. Notice that the weight of oxygen in the second case is double the weight of oxygen in the first case; i.e., 32 lb as compared to 16 lb. (See Fig. 7-4.)

If you accept the idea of atoms and molecules, you might suspect that the molecule of carbon monoxide contains only one atom of oxygen, while the molecule of carbon dioxide contains two atoms of oxygen. You would be right. The discrete manner in which elements combine to form compounds implies the atomistic nature of elements and compounds, and, therefore, implies the existence of atoms and molecules.

Take 8 lb of oxygen and 1 lb of hydrogen, somehow mix them up, and you will get 9 lb of water. Now take 8 lb of oxygen and 1.5 lb of hydrogen, somehow mix them up, and you will get 9 lb of water and 0.5 lb of unused hydrogen.* This can be understood if we assume that (1) hydrogen is made up of atoms, (2) oxygen is made up of atoms, and (3) water is made up of molecules, with each molecule having one atom of hydrogen and one atom

---

* The possibility of mixing ½ teaspoon, 1 teaspoon or 2 teaspoons of sugar in hot water seems to contradict J. Dalton. However, the resulting substance, sugar water, is not a pure compound. It is a mixture composed of sugar molecules and water molecules.

Thermodynamics

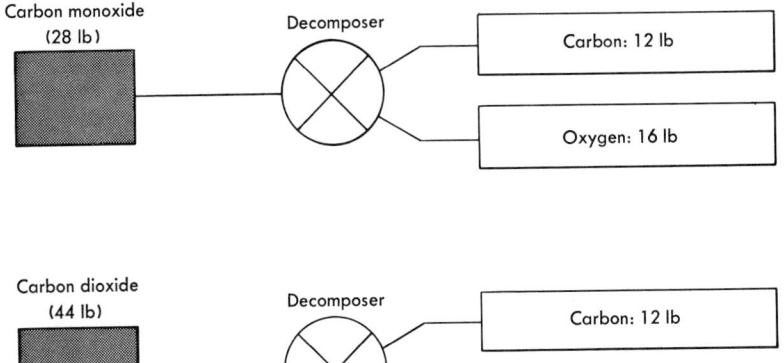

Since carbon dioxide contains twice the weight of oxygen that carbon monoxide contains, the molecule of carbon dioxide should contain twice the number of oxygen atoms that the molecule of carbon monoxide contains. Conclusion:

Carbon monoxide: C—O or CO

Carbon dioxide: $C\begin{smallmatrix}\nearrow O \\ \searrow O\end{smallmatrix}$ or CO_2

Fig. 7-4. The molecules of carbon monoxide and carbon dioxide.

of oxygen. Finally, the oxygen atom weighs eight times as much as the hydrogen atom. (See Fig. 7-5.) Of course, another possibility is that a molecule of water is composed of two oxygen atoms and one hydrogen atom; in this case, an oxygen atom would weigh four times as much as a hydrogen atom; or, as is truly the case, the molecule of water is composed of one oxygen atom and two hydrogen atoms, with the oxygen atom weighing sixteen times as much as the hydrogen atom.

All right, let's say that atoms do exist, and that the atoms of one element

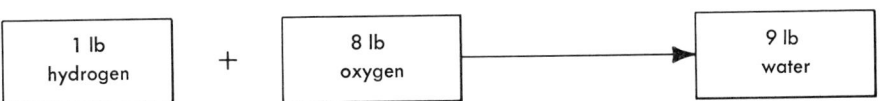

Therefore, letting H and O represent a hydrogen atom and an oxygen atom, respectively, either

$$H + O \rightarrow HO \text{ and O atom weighs } 8 \times H \text{ atom}$$

or

$$H + O\text{—}O \rightarrow HO_2 \text{ and O atom weighs } 4 \times H \text{ atom}$$

or

$$H\text{—}H + O \rightarrow H_2O \text{ and O atom weighs } 16 \times H \text{ atom} \quad \text{etc.}$$

Fig. 7-5. The possible molecular structure of water.

are different from the atoms of another element. Also that compounds are made of molecules which, in turn, are made of a collection of atoms (either the same atoms or different atoms). How, then, can we determine the exact composition of a molecule? That is to say, is a water molecule really made of one hydrogen atom and one oxygen atom, or is it made of one hydrogen atom and two oxygen atoms, or two hydrogen atoms and one oxygen atom, or . . . ?

Amedeo Avogadro of Turin, Italy

Besides the discrete manner in which the *weights* of various elements combined to form compounds, many experiments were performed in order to determine the manner in which the *volumes* of various elements combined to form compounds.

These latter experiments yielded strange results—two volumes of hydrogen combine with one volume of oxygen to form two volumes of steam.* The experimenter begins with three beakers of equal size; two of the beakers contain pure hydrogen, and one of the beakers contains pure oxygen. He combines the three beakers and gets two beakers of gaseous water ?? The experimenter began with three volumes of matter and ended with two volumes of matter.

Listen to this experiment: One volume of hydrogen and one volume of chlorine upon mixing yield two volumes of what is called hydrogen chloride gas. In this case, the experimenter begins with two volumes of matter and ends with two volumes of matter. But consider this situation: One volume of nitrogen and three volumes of hydrogen upon mixing yield only two volumes of what is called ammonia gas. In this case, the experimenter begins with four volumes of matter and ends with two volumes of matter! (See Fig. 7-6.)

Matter may be atomistic, but how do you explain this?

An Italian, Amedeo Avogadro, solved this problem in 1811 in a beautiful manner consistent with the results of Dalton. Avogadro presented two hypotheses:

1. The molecules of any gas could be composed of either two atoms, or three atoms, etc.; the atoms within these molecules could be the same, or they could be different.

> For example, one molecule of hydrogen gas could contain a combination of two atoms of hydrogen or a combination of three atoms of hydrogen, etc. (It turns out that one molecule of hydrogen gas contains two atoms of hydrogen.) While the atoms within a hydrogen molecule are the same, the atoms within a steam molecule are different. (One molecule of steam contains two hydrogen atoms and one oxygen atom.) Thus, although all of the molecules in a gas are identical, all of the atoms within the molecules may or may not be identical.

---

* All three experiments described in this section refer to the volume of the substance in the gaseous state. And, since the volume of a gas varies according to its temperature and pressure (see Sec. 8-2, Experimental Facts Concerning Gases), all of these experiments must be done at the same temperature and pressure. The experiment involving water was done at a temperature above 100°C. The other experiments can be done at room temperature.

Thermodynamics

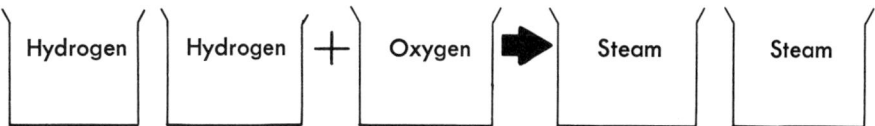

Two volumes of hydrogen plus one volume of oxygen makes two volumes of steam.

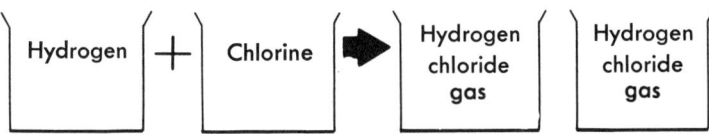

One volume of hydrogen plus one volume of chlorine makes two volumes of hydrogen chloride gas.

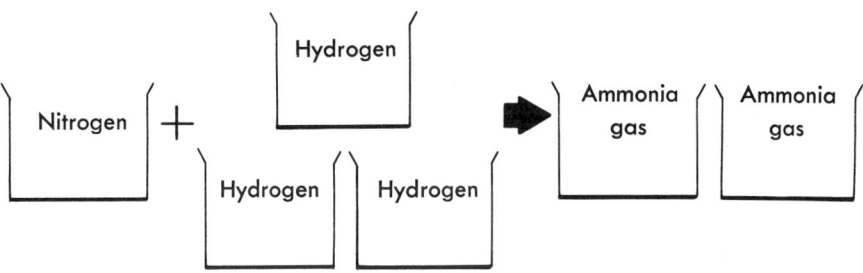

One volume of nitrogen plus three volumes of hydrogen makes two volumes of ammonia gas.

Fig. 7-6. Mixing volumes (all at the same temperature and pressure).

2. There are an equal number of molecules in every volume of a gas at the same temperature and pressure.

The second postulate states the following: Suppose that a certain volume of hydrogen gas at a certain temperature and pressure is found to contain 100 molecules. Then that particular volume, at the same temperature and pressure, could only contain 100 molecules of any gas (whether the gas be steam, hydrogen chloride, ammonia, etc.), regardless of the type, weight, or size of the molecule.

Let us see how these two postulates explained the water, hydrogen chloride-gas and ammonia-gas experiments described earlier. (See Figs. 7-7, 7-8, and 7-9.)

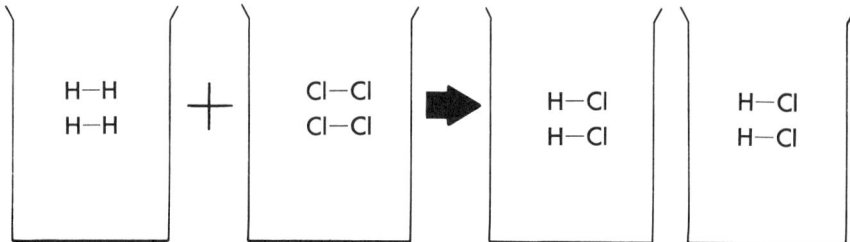

Each symbol represents 50 molecules; i.e., Cl—Cl represents 50 molecules of chlorine, each molecule having two chlorine atoms. Each volume contains 100 molecules.

Fig. 7-7. **Formation of the hydrogen chloride gas molecule (HCl).**

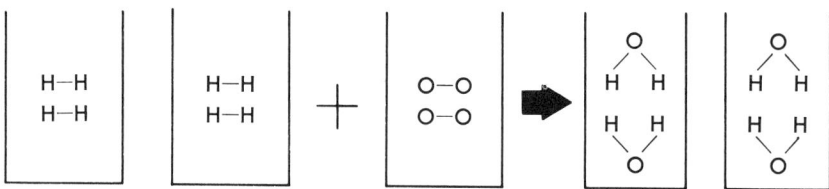

Each symbol represents 50 molecules, i.e., H—H represents 50 molecules of hydrogen, each molecule containing two hydrogen atoms. Each volume contains 100 molecules.

Fig. 7-8. **Formation of the water molecule (H_2O).**

Hydrogen chloride gas: One volume of hydrogen plus one volume of chlorine yields two volumes of hydrogen chloride gas. Let's say that, in accordance with proposition 2, the volume we use contains 100 molecules. Then 100 molecules of hydrogen plus 100 molecules of chlorine yields 200 molecules of hydrogen chloride gas. The question is, "What is the structure of the hydrogen molecule, of the chlorine molecule, and of the hydrogen chloride molecule?"

Let hydrogen be symbolized by H and chlorine by Cl. Now, the hydrogen chloride gas molecule must contain both H and Cl, and the simplest molecule is one that contains one H and one Cl. So the simplest hydrogen chloride gas molecule is HCl. There are 200 of these HCl molecules, so there must be 200 H atoms and 200 Cl atoms. But, if there are only 100 H molecules, and we need 200 H atoms, each H molecule must have two H atoms; i.e., the hydrogen molecule is H_2. Similarly, the chlorine molecule is Cl_2.

Water: Two volumes of hydrogen plus one volume of oxygen yields two volumes of water. Again, each volume contains 100 molecules. We know

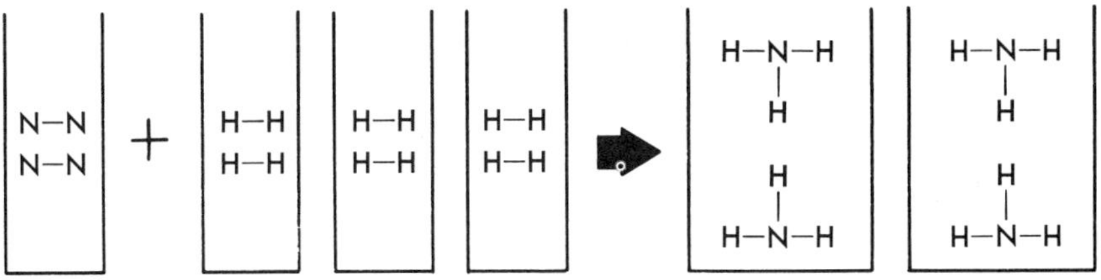

Thermodynamics

```
| N-N |   | H-H H-H |   | H-H H-H |   | H-H H-H |  ⟶  | H-N-H       H       H       H-N-H |   | H-N-H       H       H       H-N-H |
| N-N | + | H-H H-H |   | H-H H-H |   | H-H H-H |
```

Each symbol represents 50 molecules; i.e., H—N(—H above)—H represents 50 molecules of ammonia gas, each molecule containing one nitrogen atom and three hydrogen atoms. Each volume contains 100 molecules.

Fig. 7-9. Formation of the ammonia gas molecule (NH_3).

that the hydrogen molecule contains two hydrogen atoms, so the two volumes of hydrogen make a total 400 atoms of hydrogen. Our product is two volumes of water, or 200 molecules of water. Dividing the 400 atoms of H into 200 molecules of water makes 2 atoms of H in 1 molecule of water. Now water must contain oxygen as well as hydrogen, and the smallest number of oxygen atoms in a water molecule is one. So, letting O symbolize oxygen, the simplest water molecule is H_2O. Since there are two volumes of water, there are 200 water molecules, or 200 oxygen atoms. But there is only one volume of oxygen, and thus only 100 oxygen molecules. If there are 100 oxygen molecules and we need 200 oxygen atoms, each oxygen molecule must have two oxygen atoms; i.e., the oxygen molecule is O_2.

Ammonia: The student should try to prove that a nitrogen molecule is N_2 (where N is the symbol of nitrogen), and that the ammonia gas molecule is NH_3 [*Hint:* Use different colored buttons for the different atoms, and glasses for the different volumes.]

Avogadro's hypotheses were unacceptable for many decades, partly because they contradicted a basic philosophic attitude toward the elements. Elements were thought to be the simplest possible substances, and so their building blocks should also be as simple as possible. The idea of a building block of hydrogen being composed of two atoms of hydrogen is less simple than the idea of a building block of hydrogen being composed of one atom of hydrogen.

Another reason for the delayed acceptance of Avogadro's hypotheses was the image of gas particles repelling each other. A gas will occupy as large a volume as possible, and this fact promoted the idea that gas particles repelled each other. For example, if a balloon is filled with oxygen and it is punctured, the oxygen particles will race out the hole. To explain this, it was assumed that oxygen atoms in the gaseous state must repel each other. Now if oxygen atoms repel each other, how can two oxygen atoms ever combine to form an oxygen molecule? They cannot, so Avogadro must be wrong.

It wasn't until the behavior of gas particles could be understood without assuming repulsive forces, and when an enormous body of data could be easily explained by Avogadro's hypothesis, that his ideas were accepted. This occurred circa 1854.

Avogadro's Number: Definition of a Mole

How Much Does a Molecule Weigh?

Avogadro's second hypothesis leads to an interesting phenomenon. Take 10 boxes and, maintaining the same temperature and pressure in each of them, fill each one with a different kind of gas (hydrogen, oxygen, steam, helium, etc.). Each box will contain the same number of molecules. If the volume of each box is 22,400 cm^3 (22.4 liters or 0.0224 m^3) and the temperature and pressure are 0°C and 14.7 lb/in^2 respectively,* then the number of molecules in the box is called Avogadro's number.† If we know this number, and can weigh each box before and after filling, we can calculate the weight of an individual molecule. It turns out that Avogadro's number is 6.02×10^{23}.

Even without knowing Avogadro's number, we can determine the *ratio* of the weights of the molecules of the various elements by weighing all of the boxes. For example, 22.4 liters of hydrogen gas at STP has a mass of 2 g while the same volume of helium at STP has a mass of 4 g; thus, the helium molecule has twice the mass and is therefore twice as heavy as the hydrogen molecule. By definition, that quantity of a gas which occupies 22.4 liters at STP is called a *mole* of the gas. A mole of a gas contains 6.02×10^{23} molecules. The mass of one mole of a gas is the *molecular weight* of the gas.‡

7-4 STATES OF MATTER

Basically, all matter (elements, compounds, or mixtures) can exist in any one of three different states—the solid state, the liquid state, and the gaseous state. The chair on which you are sitting is matter in the solid state. The air we breathe is in the gaseous state, and the water we drink is in the liquid state. Each state corresponds to a different relationship between the molecules of the substance.

Many times the same substance can be in two different states simultane-

---

* This temperature and pressure is called *standard temperature and pressure,* and is abbreviated STP.

† The currently accepted definition of Avogadro's number is slightly, but not significantly, different from this older definition. Since the older definition is easier to picture, we will use it.

‡ Since mass and weight are different physical quantities, strictly speaking, the mass of 22.4 liters should not be called the molecular *weight*. However, history and habits are strong forces.

Thermodynamics

ously. Water can be in the solid state as well as the liquid state simultaneously; for example, ice in a glass of liquid water.

A Description of the States of Matter

The essential differences between the various states of matter are discussed below.

The Solid State

The forces between atoms or molecules in a solid are very strong, and so the atoms or molecules cannot move very freely. They tend to arrange themselves into a lattice framework. Each atom or molecule is fixed at one lattice site, and it can only vibrate around this site. (See Fig. 7-10.)

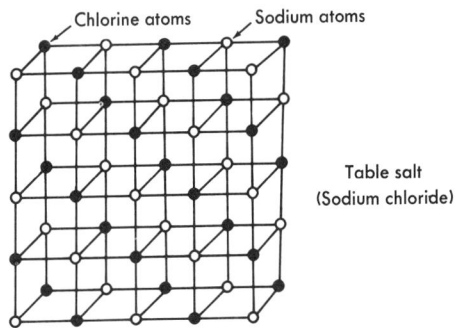

Table salt (Sodium chloride)

Molecules (in this case, atoms) are constrained to a lattice framework and can only vibrate about their lattice site.

Fig. 7-10. Microscopic view of the solid state.

The Gaseous State

The atoms or molecules in the gaseous state are independent of each other. They are free to move as they please, since there is no force* being exerted on one by any of the others. The only time one atom or molecule affects another is during a collision. They collide and bounce away. The speeds at which they travel are enormous by everyday standards, about 500 m/sec (1640 ft/sec). A car going 60 mi/hr is traveling at 88 ft/sec, or less than 10 per cent of the speed of a gas molecule. (See Fig. 7-11.)

The Liquid State

The size of the forces between atoms or molecules in a liquid lies between those of a solid and a gas. The forces are not strong enough to insist that

---

* In reality, there are very slight forces between molecules in a gas, but for our purposes they need not be considered.

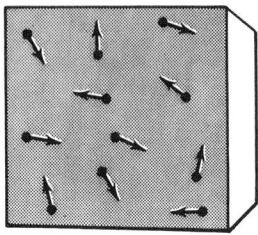

Fig. 7-11. **Microscopic view of the gaseous state.**

Molecules of a gas are moving freely in a closed container.

each atom or molecule remain fixed at one lattice site, nor are the forces so weak that every atom or molecule is independent of each other. (See Fig. 7-12.)

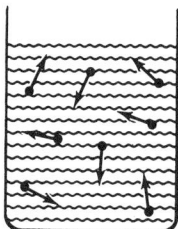

Molecules of a liquid are not completely free, but they are not completely constrained either.

Fig. 7-12. **Microscopic view of the liquid state.**

Characteristics Which Determine the States of Matter

What determines the state in which a particular type of matter will exist? Its particular *molecular composition,* the *pressure* on its surface, and its *temperature.* Let's consider each in turn.

Molecular Composition

If the substance is a compound, the peculiarities of the atoms which compose the molecules of the substance, as well as the spatial configuration of the atoms within the molecule, help determine the state of the substance. For example, the peculiarities of hydrogen and oxygen atoms, as well as the spatial configuration of these atoms* within the water molecule, help determine the state of water. If the substance is a pure element; i.e., gold, or silver, the peculiarities of the atoms of that element help determine the state of the substance.

---

* The way in which the spatial configuration of the atoms within the molecules determine the state of the compound is quite complex, and will not be discussed in this book.

Thermodynamics

Pressure

The effect of pressure on the state of the substance will be deferred until page 117 and even then, it will be discussed only briefly.

Temperature

The single, most important item in determining the state of a substance is the temperature of that substance. (See Fig. 7-13.) At 0°C, water goes from the liquid to the solid state; at 100°C, water goes from the liquid to the gaseous state.

Temperature is a fairly subtle concept and like all subtle concepts, it is easier to talk about its measurement than its essence.* So let us talk about its measurement.

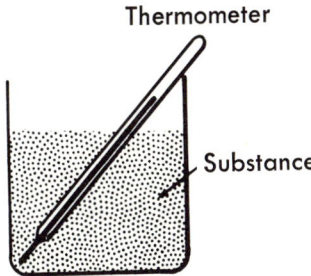

The temperature of a substance is extremely important in determining the state of the substance.

Fig. 7-13. Temperature and the state of a substance.

7-5 TEMPERATURE

The Beginning of Thermometry

Temperature is not a new word. We all have a sense of hotness or coldness, and can almost always tell which of two bodies is at a higher temperature. However, because each of us has a different sensitivity to hot and cold, and because we are human beings and so easily fooled, it would be much better to assign numbers to various degrees of hotness and coldness and have a definite way of measuring these numbers. Then, different people at different laboratories would always understand each other and would be able to repeat experiments accurately.

---

* Time is a perfect example of a concept whose essence is mysterious, but whose measurement is trivial.

The Galilean Thermometer

One of the first known attempts at numbering temperature is credited to Galileo in approximately 1592. Galileo knew that gases expand or contract as the temperature rises or falls so, because air is a gas, air should expand or contract when the temperature rises or falls. Galileo thought of using the change in volume of air to assign numbers to various temperatures.

He poured colored water into a spherical glass ball with a long stem until it was about half full. He then inverted the device into a flask of water. There will now be colored water in the stem of this thermometer.* The remainder of the thermometer (spherical ball and part of the stem) will contain air. If a mark is placed at the line separating the water and air, that line can be called "room temperature," or "5° Galileo," or whatever. (See Fig. 7-14.)

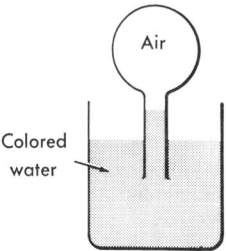

As the temperature of the air rises or falls, the air expands or contracts causing the boundary between the colored water and the air to fall or rise.

Fig. 7-14. **The Galilean thermometer.**

Now, to measure the temperature of a body, place the spherical ball in contact with the body. If the body is at a higher temperature than the air in the spherical ball, the air in the spherical ball will expand, pushing the colored water farther down the stem of the thermometer. The line between the water and air is lower than the previous line was, and a new mark can be made on the stem indicating the temperature of that particular body. This new mark is assigned a number also. Conversely, if the body is at a lower temperature than the air in the spherical ball, the air in the spherical ball will contract, and water will climb up the stem of the thermometer. The line between water and air moves up, and a mark and number associated with this new temperature is made. In both cases the air inside the spherical ball reaches the temperature of the object under study, and then either expands or contracts depending upon whether the body is at a higher or lower temture than the air inside the thermometer.

---

* A thermometer is the name of a device that can assign numbers to various degrees of hotness and coldness. Galileo used colored water because its boundary is more easily seen.

Thermodynamics 117

There are two problems stemming from the above remarks.

1. What kind of thermometer should we use? Is Galileo's air thermometer satisfactory, or would another type of thermometer be more accurate and reproducible? A major problem with Galileo's thermometer lay in the fact that variations in atmospheric pressure affected the height of the water in the stem. The solution to this problem led to the development of the mercury thermometer.

2. How should we assign numbers corresponding to the various temperatures? The solution to this problem led to the adoption of the Fahrenheit and centigrade temperature scales.

Let's consider each problem separately.

Thermometers and Temperature Measurement

Development of the Mercury Thermometer

During the time interval 1592 to 1750, the Galilean air thermometer was replaced by a mercury thermometer. The development occurred as follows:

Since liquids expand or contract as the temperature rises or falls, it is possible to use the change in volume of liquids as well as that in gases to measure temperature. Around 1650, wine was poured into a sealed glass tube (no air was in the tube), and the expansion or contraction of this liquid was used to measure temperature. (Wine was used rather than ordinary water, probably because wine is naturally colored and is therefore easier to see.) But if things got too cold, the wine would freeze and if things got too hot, the wine would boil. (Besides that, wine is for drinking.) So what is needed is a liquid that remains a liquid over a wide range of temperatures. Mercury satisfies that requirement. (See Fig. 7-15.)

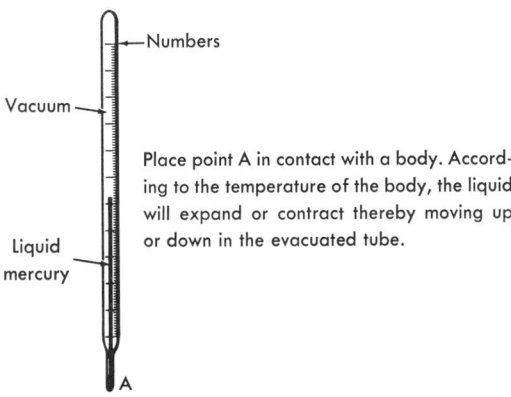

Fig. 7-15. **The mercury thermometer.**

The use of mercury was first introduced in 1659, but there were problems associated with impurities in the liquid and techniques of filling the glass tube, etc. Finally, in 1724, Gabriel Daniel Fahrenheit published a number of articles describing a technique for making fine, reproducible, mercury thermometers. He also described how he used these thermometers to assign numbers to two different states of water.

Degrees Fahrenheit, Degrees Centigrade

5°, 88°, 72°, etc. What is all this?

Instead of assigning numbers to various temperatures in a random fashion, it is really more meaningful to assign numbers to just two different states of the same substance (one state being hot, the other being cold) and then divide the resulting interval into a certain number of parts or divisions. For example, if state A is very cold and state B is very hot and there are twenty divisions between them, one division would mean 1/20 of the temperature between A and B. Notice that if only one state and one number had been chosen, it would not be clear how large a difference in temperature corresponded to one division. For example, if only A is used and it is assigned the number 10, to what would the number 11 correspond? It is 10 per cent more than A, so perhaps it is 10 per cent hotter? But I could have just as easily assigned the number 52 to state A. Then what would 53 mean? By assigning two numbers between two different states, the size of a division is fixed.

However, the choice of what particular substance was to be used, what particular states of that substance, what numbers to assign those states, and how many divisions should lie between those two states was still arbitrary.

Since it is desirable for all experimenters to be able to understand each other, the kind of substance used should be easily available, and the states of that substance should be easily reproducible. In this way, all experimenters can check each other's work and mean the same thing when they say the same thing; i.e., 10° in New York is 10° in Hong Kong, San Francisco, and/or Paris.

After a good deal of trial and error, pure water was the agreed-upon substance. In 1724, Fahrenheit presented the number 32 degrees Fahrenheit (or 32°F) for melting ice (i.e., a mixture of ice and water), and 212°F for boiling water (i.e., a mixture of water and steam).* (See Fig. 7-16.) Notice that melting ice is a mixture of solid and liquid water, while boiling water is a mixture of liquid and gaseous water.

If we recall that pressure as well as temperature affects the state of matter, we will realize that the above numbers are valid only at standard atmospheric pressure ($14.7 \text{ lb}/\text{in}^2$ or $10^5 \text{ nt}/\text{m}^2$). If the pressure is different,

---

* These peculiar numbers were convenient to Fahrenheit at the time. The choice of pure water as the substance, as well as the choice of melting ice and/or boiling water as the states to which numbers should be assigned, was actually suggested in 1665 by Huygens. But, I guess, like most suggestions, it required the proper timing, the proper person, and the proper atmosphere of thought in order to be utilized. The suggestion lay at rest for decades waiting proper instruments and properly directed researchers.

Thermodynamics

Ice-water
Temperature: 32° F
or 0° C

Boiling water
Temperature: 212° F
or 100° C

Fig. 7-16. The definition of the Fahrenheit and centigrade temperature scales.

the temperatures are different; i.e., a thermometer that reads 32°F when placed in a mixture of ice and water at sea level will read about 25 or 30° when the mixture is placed on a mountain top; also, if water is heated on a mountain top, it will not boil at 212°F, but probably at 180 or 200°F.

Around 1750, some scientists introduced a new temperature scale and a new set of numbers; namely, the centigrade scale, with the number 0°C for melting ice, and 100°C for boiling water. The centigrade scale is very popular in the scientific world. We will use the centigrade scale.

If you took some ice from your refrigerator, mixed it with water, and then measured the resulting temperature, I, for one, would be surprised if you read 0°C. That is because tap water is generally very impure. It contains various sorts of atoms other than hydrogen and oxygen, and this tends to distort the temperature. The effect is slight, only 1° or 2°C, but it is there. The same problem occurs with boiling water. The impurities mask the true temperature.

Absolute Zero, the Kelvin Temperature Scale

Although liquid thermometers (particularly mercury thermometers) proved much more convenient than gaseous thermometers, the behavior of gaseous thermometers led to a very interesting observation. It was found that the volume of a gas increased linearly with temperature. That is, if you increase the temperature of a gas by 10°C, the volume increases from 1.000

to 1.037 units. And if you increase the temperature by 20°C, the volume will increase from 1.000 to 1.074 units. There is a .037-unit increase in volume for every 10°C increase in temperature. A graph illustrating this can be drawn, with the volume of the gas on the vertical axis, and the temperature of the gas on the horizontal axis. (See Fig. 7-17.)

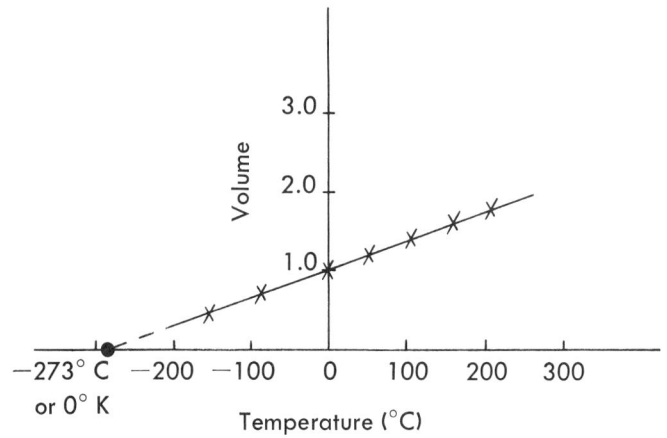

Fig. 7-17. **The determination of absolute zero.**

The crosses correspond to the volume of the particular gas at various temperatures. At 0°C, volume is 1.000 units, at 50°C volume is 1.185 units, etc. At −50°C, the volume is 0.815, and at −100° the volume is 0.630 units.

The volume continues to decrease as the temperature gets colder and colder. Is it possible to get this amount of gas so cold that it would occupy zero volume?

Of course not! The gas will become a liquid before it will occupy zero volume.

But just suppose that it were possible to cool this gas down to zero volume. How cold would that be? By continuing the straight line to zero volume and reading the temperature indicated at that point, one could find a measure of that coldness. The temperature is −273°C.* It is called *absolute zero*. This experiment can be done with any kind of gas, or any amount of gas. The temperature at which the gas would occupy zero volume is always −273°C.

A new temperature scale was introduced which called that point 0° Absolute or 0° Kelvin (0°K). This was in honor of a famous physicist (c. 1900) William Thomson who, in 1892, was given the title Baron Kelvin. The temperature of the ice-liquid water combination is 273°K (corresponding to 0°C) and the temperature of boiling water is 373°K (corresponding to 100°C). Notice that, because the interval between melting ice and boiling water is 100°K, as well as 100°C, the size of one

---

* The absolute zero point is actually −273.16°C, but we shall disregard the .16° difference.

Thermodynamics

degree Kelvin is the same as the size of one degree centigrade; it is just displaced by 273 degrees. Unless otherwise specified, any formula which contains temperature in it, means temperature in degrees Kelvin.

7-6 HEAT AND TEMPERATURE

Is heat the same as temperature?

No! In everyday life we so often couple heat and temperature that it is difficult to reconcile the fact that they are entirely different entities.

Heat is a form of energy, while temperature is a degree of hotness and coldness that can be measured on a definite scale and is related to the direction of energy flow. For example, if two bodies are at different temperatures, energy will flow from the higher temperature body to the lower temperature body.* The rate of energy flow depends on the difference in temperature between the two bodies. If both bodies are at the same temperature, no energy flow is possible. Therefore, the concept of temperature can be looked upon as a way of describing whether or not energy will flow between two bodies, and the numerical measurement of temperature helps determine the rate of energy flow. (See Fig. 7-18.)

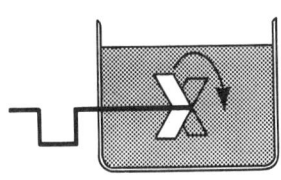

Heat is a form of energy.
By heating the water with a flame, I can increase its temperature by 1° C.
By turning a paddle wheel, I can increase its temperature by 1° C.

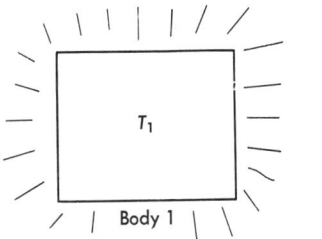
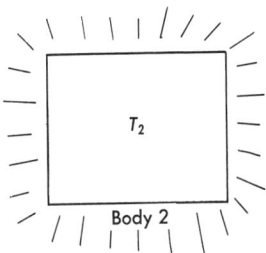

If T_1 is greater than T_2, there will be a net energy transfer from body 1 to body 2.
If T_1 equals T_2, there will be no net energy transfer.
If T_1 is less than T_2, there will be a net energy transfer from body 2 to body 1.

Fig. 7-18. **Heat is a form of energy. Temperatures indicate the direction of energy flow.**

---

* The form of energy that flows between two bodies as a result of their temperature differences is radiant energy.

7-7 REVIEW

There are 103 different kinds of elements, and the building block of each element is an atom; therefore, there are 103 different kinds of atoms. If two or more atoms combine, the resulting building block is called a *molecule*. A substance is composed of molecules. If all of the atoms within the molecule of a substance are identical, that substance is an element, but if the atoms within the molecule of a substance are different, then the substance is either a compound or a mixture. If all of the molecules of a substance are identical, then the substance is a compound, but if all of the molecules of a substance are not identical, the substance is a mixture. For example,

Hydrogen gas	Molecule: H_2	An element
Water	Molecule: H_2O	A compound
Sugar water	Molecule: (a) H_2O (b) sugar molecule	A mixture

All matter can exist in three different states: solid, liquid, or gas. Matter, which is homogeneous, is called a substance and the major characteristic of a substance which determines whether it will be solid, a liquid, or a gas is the temperature of the substance. The temperature of a substance can be measured with a thermometer. There are three temperature scales in common use: the Fahrenheit scale, the centigrade scale, and the Kelvin scale. (The centigrade temperature scale is very similar to the Kelvin temperature scale, except that the latter scale starts from absolute zero.)

Although we tend to equate heat and temperature, they are quite different from each other. Heat is a form of energy while temperature is a degree of hotness or coldness that can be measured on a definite scale, and is related to the direction of energy flow between two bodies.

In the next chapter, we will see how the temperature of a gas can be understood by relating it to the average kinetic energy of the molecules in the gas. Unfortunately, liquids and solids are more complicated, and so the temperature of a liquid or a solid is not understood as readily as the temperature of a gas.

PROBLEMS

1. What characteristics differentiate one substance from another?
2. Sketch the evidence that implied that matter was atomistic.
3. Sketch the evidence that led to discovering the molecular structure of the elements.
4. Is it possible to satisfy the hydrogen plus oxygen volume experiments by saying that a molecule of hydrogen is H_4 and that a molecule of water is H_4O?
5. If 22.4 liters of hydrogen gas at STP has a mass of 2 g, what is the mass of a single hydrogen molecule? (Avogadro's number is 6.02×10^{23} molecules/mole.)

Thermodynamics

6. Describe one problem associated with Galileo's thermometer.
7. How does the Kelvin temperature scale differ from the centigrade temperature scale?
8. Is heat only another word for temperature? Explain.
9. If two bodies are at the same temperature, is any energy flow between the two bodies possible?

Chapter 8

The Kinetic Theory of Gases

8-1 INTRODUCTION

This chapter is divided into three parts. The first part (Sec. 8-2) discusses various experimental facts concerning gases. The next part (Sec. 8-3) presents the so-called kinetic theory of gases. This theory is based on a particular model of a gas; i.e., assumptions are made concerning the size, nature, and behavior of the molecules in a gas and these assumptions constitute our model of a gas. A prediction relating the pressure and volume of the gas to the average kinetic energy of the molecules in the gas is deduced from this theory. Finally, the last part of this chapter (Sec. 8-4) shows that if we accept one reasonable relationship; namely, the relationship between the temperature of a gas and the average kinetic energy of a molecule in this gas, then the theory deduced from this particular model of a gas corresponds exactly to the experimental facts.

The chapter ends with a warning concerning the reality of a model versus the reality of reality.

8-2 EXPERIMENTAL FACTS CONCERNING GASES

The Perfect Gas Law

Take a cylinder, fill it with a mole* of a gas and place a piston on top so that the gas is completely enclosed. Also, make provisions for measuring the

---

* A mole of a gas is that quantity of a gas that occupies 22,400 cm^3 at a pressure of 14.7 lb/in^2 and at a temperature of 0°C. A mole of a gas contains 6.02×10^{23} molecules. See Sec. 7-3.

The Kinetic Theory of Gases

volume and temperature of the enclosed gas.* The pressure of the gas is determined by the weights placed on the piston; i.e., the pressure of the gas is the net weight on the piston divided by the area of the piston.† (See Fig. 8-1.)

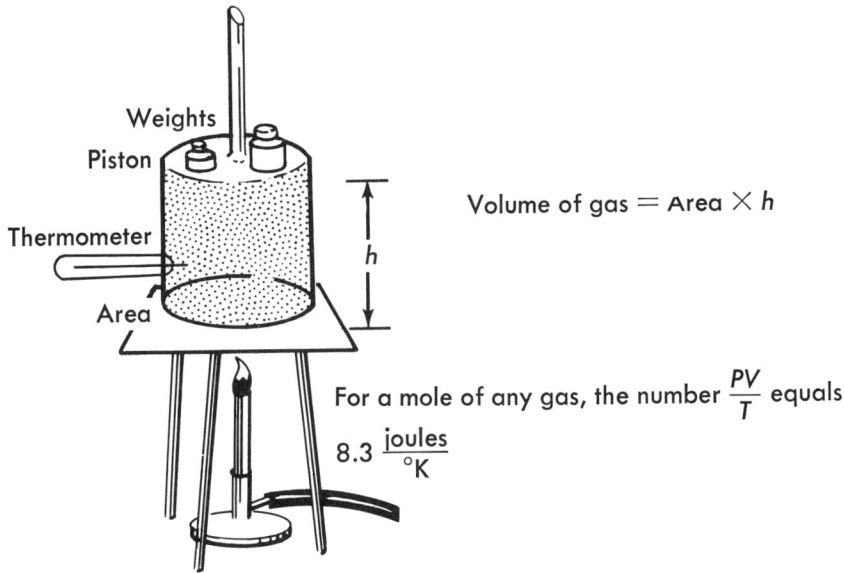

Volume of gas = Area × h

For a mole of any gas, the number $\frac{PV}{T}$ equals 8.3 $\frac{\text{joules}}{°K}$

Fig. 8-1. Experiment to determine pressure, volume, and temperature relationships in a gas.

Now let's begin our experiment.

We can vary the pressure of the gas by changing the size of the weights. We can also vary the temperature of the gas by placing our cylinder above a small flame. Now there are many different equilibrium states, and every time we reach equilibrium we compute the pressure, volume, and temperature of the gas. Take the value for the pressure (P), multiply it by the volume (V), and divide the product by the temperature (T) in °K. Thus, if

$$P = 50 \frac{\text{nt}}{\text{m}^2}, V = 3 \text{ m}^3, \text{ and } T = 18.1°K,$$

then

$$\frac{PV}{T} = \frac{50 \times 3}{18.1} = 8.3 \frac{\text{joules}}{°K}$$

---

* To measure volume, the cylinder walls can be made transparent and markings can be used to indicate the height of the gas in the cylinder. A thermometer can be inserted in a suitably designed hole in the cylinder wall to measure temperature.

† Weights are placed on the piston and the piston falls. When the piston stops falling, the pressure that the piston is exerting on the gas is equal to the pressure that the gas is exerting on the piston.

If we purposely change the pressure, the volume and the temperature will change, so that PV/T remains the same. Similarly, if we purposely change the temperature, the pressure and volume will change, so that PV/T remains the same. The surprising thing is that over an extremely wide range of pressure, volume, and temperature, the number PV/T remains the same.

Not only is PV/T constant for any one gas, but the number is the same for a mole of any gas. That is, if we take a mole of any one gas, the number PV/T for that one gas is the same as the number PV/T for a mole of any other gas. The number PV/T is independent of the type of gas in our cylinder. That number is 8.3 joules/°K (in MKS units). This fact is known as the *perfect gas law*.

Like all beautiful things, the perfect gas law has its imperfections. Although the law is valid for an extremely wide range of pressure, volume, and temperature, it is not valid for all pressures, volumes, and temperatures. Furthermore, if extremely accurate measurements of P, V, and T are made, it is found that not all gases behave in quite the same way. Thus, within a finite (not infinite) range of pressure, volume, and temperature, and within a limited (not unlimited) accuracy, the perfect gas law is true.

However, we can imagine the existence of an *ideal gas* for which the perfect gas law is valid with unlimited accuracy for all P, V, and T. The value of PV/T for one mole of an *ideal gas* is 8.3 joules/°K.

Development of the Perfect Gas Law

The perfect gas law was not recognized suddenly. At first, in 1662, Robert Boyle noticed that, at a constant temperature, the product of pressure times volume for a given quantity of gas was constant. Then around 1800, it was noticed that at a constant pressure, the quotient of volume divided by temperature for a given quantity of gas was constant.

Boyle's Law

In 1662, Robert Boyle, an Englishman, noticed that for a given quantity of gas, the pressure varies inversely with the volume of the gas, provided that the temperature remained constant. Mathematically,

$$P = \frac{K_1}{V} \quad \text{or} \quad PV = K_1 \quad \quad (8\text{-}1)$$

where K_1 is some number depending on the quantity of gas and the temperature.

Charles's Law

Around 1800, J. A. C. Charles, a Frenchman, noticed that for a given quantity of gas, the volume varied directly with the temperature of the gas, provided that the pressure remained constant. Mathematically,

The Kinetic Theory of Gases

$$V = K_2 T \quad \text{or} \quad \frac{V}{T} = K_2 \tag{8-2}$$

where K_2 is some number depending on the quantity of gas and the pressure.

Now if, for a given quantity of gas, we allow the pressure, volume, and temperature to vary, what will happen? Combining Boyle's Law and Charles's Law, we have

$$\frac{PV}{T} = K_0 \tag{8-3}$$

where K_0 is some number which depends on the quantity of gas we have chosen.

Does the PV/T change as we go from gas A to gas B? Or is PV/T the same for gas A as it is for gas B?

If the mass of gas A equals the mass of gas B, then the PV/T for gas A is different from the PV/T for gas B. If, however, the number of molecules in gas A equals the number of molecules in gas B, then the PV/T for gas A equals the PV/T for gas B. It is the *number of molecules* that is important in the PV/T relationship.

Because a mole of a gas always contains 6.02×10^{23} molecules, PV/T for a mole of one gas is the same as PV/T for a mole of any other gas. The value of PV/T for any one mole of any gas is 8.3 joules/°K.

8-3 THE KINETIC THEORY OF GASES

Model of an Ideal Gas

In an attempt to understand why PV/T is a constant for all gases, scientists concocted a model of an ideal gas. (See Fig. 8-2.) The assumptions concerning the ideal gas were as follows:

1. All of the molecules of our ideal gas have the same mass and are moving about in a random fashion; i.e., there are as many molecules moving in any one direction as in any other direction.
2. The molecules are quite small, and the distance between them is quite large.
3. There is no force between the molecules, and the only time they affect one another is when they collide.
4. No kinetic energy is lost in any collision between a molecule and the wall, or between one molecule and another molecule.
5. The gravitational force of the earth on the molecules is assumed to have a negligible effect on the motion of the molecules.*

---

* The molecules move so rapidly that they collide with the walls or with each other before gravity can appreciably affect their motion.

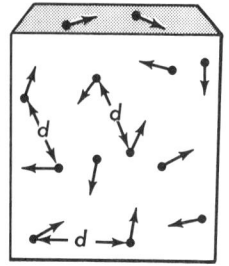

d is large compared to size of molecule.

All molecules have the same mass, and all molecules move in a random fashion.

There is no intra-molecular force.

No kinetic energy is lost during collisions. Molecules move too rapidly to be affected by gravity.

Fig. 8-2.   An ideal gas.

Deductions From the Model

Now, the problem is to try to predict the relationship between P, V, and T in such an ideal gas.

Let's assume that we have a cubic box of side L, and just one molecule racing back and forth along the x-direction. (See Fig. 8-3.) Every time the molecule hits the wall, it exerts a certain force on the wall. This force

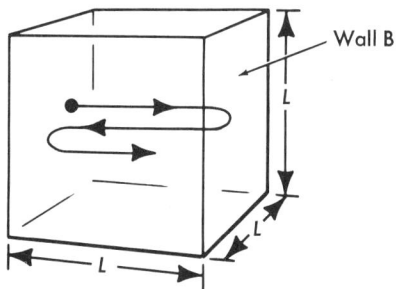

A single gas molecule, mass m, velocity v, moving back and forth along the x-axis. Time between collisions at wall B is $2L/v$.

Fig. 8-3.   Pressure on wall B due to a single gas molecule.

divided by the area of the wall is the pressure exerted on the wall by the molecule. Mathematically,

$$P = \frac{F}{A}$$

where
$\quad\quad\quad P$ is the pressure,
$\quad\quad\quad F$ is the force, and
$\quad\quad\quad A$ is the area of the wall.

The Kinetic Theory of Gases

Now,
$$F = \frac{\text{change in momentum*}}{\text{change in time}}$$

$$F = \frac{\text{Momentum}_{\text{final}} - \text{Momentum}_{\text{initial}}}{\text{Time between collisions}}$$

Using assumption #4, however, we can show that

Momentum$_{\text{final}}$ = mv and
Momentum$_{\text{initial}}$ = $-mv$

and from the geometry of the box, we have

Time between collisions = $\frac{2L}{v}$.

So,
$$F = \frac{(mv) - (-mv)}{\frac{2L}{v}}$$

$$= \frac{2mv}{1} \times \frac{v}{2L}$$

$$F = \frac{mv^2}{L}$$

Since A is L^2,
$$P = \frac{F}{A} = \frac{mv^2}{L^3}$$

But L^3 is the volume of the box, so
$$P = \frac{V}{mv^2}$$

or
$$PV = mv^2$$

Thus, for a *single* molecule moving in the x-direction, the pressure times the volume equals mv^2. But mv^2 is $2 \times \frac{1}{2} mv^2$, or $2 \times$ the kinetic energy of the molecule. So we can write

$$PV = 2 \times \text{(K.E.)} \tag{8-4}$$

where (K.E.) is the kinetic energy of the molecule.

If many molecules are placed in the same box, the pressure is of course increased. But how many of these molecules can be considered to be moving in the x-direction? Since there are three independent directions in space (x, y, and z), only $\frac{1}{3}$ of the given number of molecules in the box can really be considered to be moving in the x-direction.† (See Fig. 8-4.)

---

* This expression for the force can be shown to be exactly equal to $F = ma$.
† This deduction depends directly on assumptions 1 and 5. We can interpret assumptions 2 and 3 as saying that each of the N molecules acts as if none of the other molecules existed; i.e., each molecule acts independently.

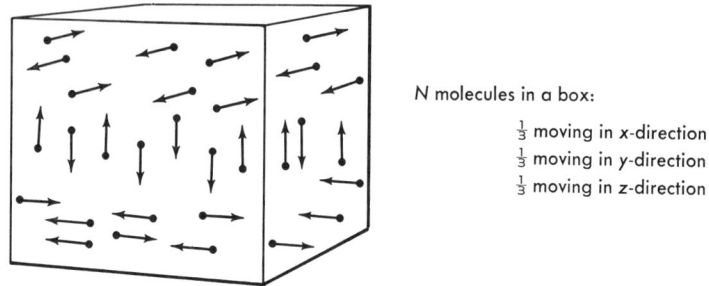

Fig. 8-4. Schematic diagram of *N* molecules in a box.

Mathematically,*

$$PV = \tfrac{1}{3} N \times 2 \text{ (K.E.)}$$

or

$$PV = \tfrac{2}{3} N \times \text{(K.E.)} \tag{8-5}$$

where N is the number of molecules in our box.

8-4 AN INTERPRETATION OF THE TEMPERATURE OF A GAS

If we say that the temperature of a gas is related to the kinetic energy of a gas molecule, then K.E. $= bT$, where b is the constant relating the K.E. of a gas molecule to the temperature. The number b is called *Boltzmann's constant*. But now we can write Eq. (8-5) above as

$$\begin{aligned} PV &= \tfrac{2}{3} N \text{ (K.E.)} \\ &= \tfrac{2}{3} N \, bT \end{aligned}$$

or

$$\frac{PV}{T} = \tfrac{2}{3} Nb \tag{8-6}$$

Notice that PV/T is related to the number of particles in our box and to the proportionality constant between kinetic energy and temperature. But since the latter is a constant of nature and does not depend on the type of molecule at all, PV/T, for our model of an ideal gas, depends only on the *number* of molecules in our box and not on the *type* of molecules involved. This is the essence of the perfect gas law that was presented earlier in Sec. 8-2. If we concern ourselves with one mole of a gas, then $N = 6.02 \times 10^{23}$. We can determine the value of b from other measurements, and so we can calculate that $\tfrac{2}{3} Nb = 8.3$ joules/°K. So, PV/T for one mole of an ideal gas is 8.3 joules/°K.

---

* I am taking the liberty of assuming that all of the molecules have the same velocity and therefore the same kinetic energy. Strictly speaking, I should use average velocities and average kinetic energies, but with this simplification, the mathematics is easier and the physics is just as accurate.

The Kinetic Theory of Gases

8-5 REVIEW

We began with an experimental fact that is almost always true; namely, that $PV/T = 8.3$ joules/°K for one mole of any gas. We then adopted a model of an ideal gas and a set of assumptions concerning the molecules of this gas. Using this model of a gas, we derived an expression which stated that $PV/T = 8.3$ joules/°K for one mole of any gas. This prediction is exactly identical to the experimental fact concerning gases.

Does this verification mean that our model of a gas is true? Does this mean that the molecules in a gas really do obey assumptions 1, 2, 3, 4, and 5? No! Not quite!

The reality is that PV/T is a constant for all gases within a limited range of accuracy and within a finite range of pressure, volume, and temperature. The model we have adopted allowed us to understand why this phenomenon *may* be so. A model is a thinking tool which may or may not have anything to do with reality. In this case, the model yielded fairly accurate predictions, so, while our assumptions may not be completely true, they probably contain some truth.

The model contains some well-known mistakes. For example, I assumed that a gas molecule went back and forth, from wall to wall, and did not collide with any other molecule. But we know that normally a molecule travels less than 0.01 m before it collides with another molecule. Also, assumption 3 states that there are no forces between molecules in a gas. However, there are forces between the molecules in a gas. It is true that the forces are small, but they do exist. In order to eliminate these mistakes as well as some others, modifications to the model can be made. And if the modifications are not too extreme, our mathematics can cope with the problem and accurate predictions are obtained. Sometimes, however, the modifications required are too extreme and our mathematics become impotent. In that case, approximations are made.

We simply cannot escape the fact that nature is within a hair's breadth of being as complicated as a human being.

In spite of the faults of our model, it was highly successful. In a way, it was too successful because it enhanced the mechanistic view of the universe. This view looked upon the universe as a huge machine with nothing but motion, particles, and forces. The mechanistic view was very deterministic; i.e., according to Newton's laws of motion, the future is precisely determined by the present, and the present is precisely determined by the past. According to the mechanistic view, if we know what all of the particles in the universe are doing now, we can determine exactly what they will be doing at any future time. All we need is the diligent application of Newton's laws of motion, and lots of paper and pencils. Needless to say, this mechanistic viewpoint delayed the understanding and acceptance of many modern concepts of space, time, and matter. Perhaps you will study the more modern concepts later in your course.

PROBLEMS

1. One mole of hydrogen gas is enclosed in a container whose volume is 2m³ and whose temperature is 27°C (300°K). What is the pressure on the walls of the container?
2. If the temperature of a gas remains constant, and its volume doubles, what happens to the pressure?
3. If the volume of a gas remains constant, and its temperature doubles, what happens to the pressure?
4. A gas is enclosed in a cube whose sides are 1 meter in length. The gas is at room temperature (27°C or 300°K) and a gas molecule at room temperature has a velocity of about 500 m/sec (1100/mph). How long does it take a gas molecule to travel from one wall to another?
5. Due to gravitation, the gas molecule in Problem 4 will accelerate downward. Using the result of Problem 4, compute the distance a gas molecule falls as it goes from one wall to another. (See figure below.) In the light of this result, is assumption 5 reasonable?

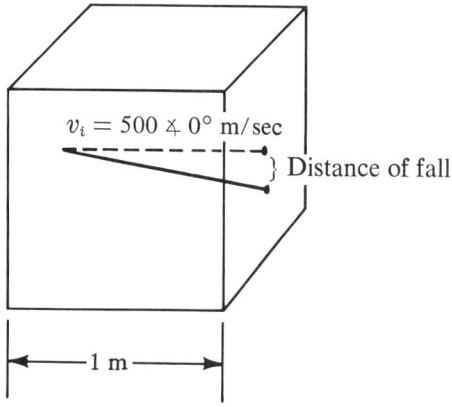

Chapter 9

Sound

9-1 INTRODUCTION

A starving bat can detect, chase, and swallow a mosquito once every six seconds; in pitch darkness, that is. Apparently, bats emit high-frequency sound waves and interpret the reflection pattern, or echo pattern, that results from the collision of these sound waves with the poor mosquitoes. Sound waves can exist in an enormous variety of frequencies; however, the human ear is only sensitive to a limited region of this frequency spectrum. Generally speaking, the frequencies emitted by bats are too high for us to hear.

But we are not deprived of the joy of sound. The sound of a human voice can provide either warmth and understanding, or coolness and indifference. Indeed, from the crying of a newborn child to the music of a 100-piece orchestra, sound fills our lives.

What exactly is sound?

Sound is a certain kind of wave, and in order to understand and appreciate the phenomena of sound we need to develop a vocabulary. Each word in the vocabulary describes a certain feature of waves in general, and of sound in particular.

9-2 WAVES

Vocabulary and Characteristics

Suppose that I have a very long rope, and I attach one end of it to a very distant wall. I then continually raise and lower the free end a distance $+A$ meters up and $-A$ meters down. (See Fig. 9-1.) Basically, two things are happening:

Sound
a certain kind of wave

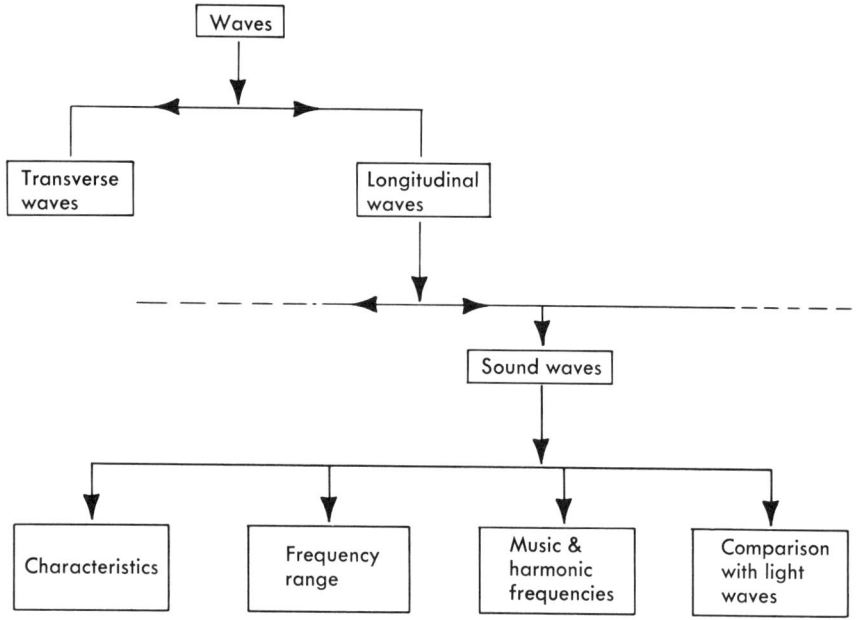

Sound

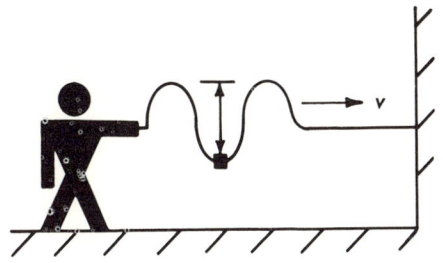

Each segment moves up and down while the wave moves toward the right.

A rope-wave being produced by the author.

Fig. 9-1.   Example of a transverse wave.

1. A wave is traveling from the rope toward the wall with a certain velocity.

Notice that although the wave is moving toward the wall, each segment of the rope is simply moving up and down. That is to say, the average displacement of any one segment of the rope is zero, but the wave, which is composed of all the segments of the rope, is moving toward the wall. The wave represents a certain amount of energy and momentum that is being transmitted by me toward the wall. I could transmit energy and momentum via a particle (by throwing the particle), or I could produce a wave and transmit them via this wave.

2. Every segment of the rope is undergoing periodic motion; i.e., every segment of the rope is going up and down.

The time for one complete cycle is called the *period* (T), and the number of cycles per second is called the *frequency* (f). The frequency is the inverse of the period; i.e., $f = 1/T$. Also, each segment of rope has a certain *amplitude* (A). The amplitude is defined as the maximum displacement from the equilibrium position.

I am holding the initial segment of the rope, and I can move it high above its equilibrium position or I can move it only slightly above its equilibrium position. If I raise this segment 0.1 meter, 0.5 meter, or 1.0 meter, I give the wave an amplitude of 0.1 m, 0.5 m, or 1.0 m, respectively. The amplitude of the wave, then, is determined by myself, the source.

The frequency of the wave is also determined by myself, the source. The frequency of the wave is the same as the frequency of the segments of rope composing the wave. I am holding the initial segment of rope, and if I move this segment up and down quickly, the frequency of all of the segments is higher than it is if I move this segment up and down slowly. That is, I can take 1 day, 1.0 second, or 0.1 second to move the initial segment of the rope up and down. All of the segments of the rope, and therefore, the wave, would then have a frequency of 1/86,400 cps, 1/1 cps, or 1/0.1 cps, respectively. (There are 86,400 seconds in one day.)

If the amplitude and frequency of the wave are determined by the source, what determines the velocity of the wave?

The velocity of the wave is determined by the medium in which the wave is traveling; i.e., by the rope itself. Both the tension in the rope and the mass of the rope affect the velocity of the wave. As the tension (or stiffness) of the rope increases, the velocity increases. But as the mass of the rope increases, the velocity decreases. A tauter rope then produces a greater wave velocity, while a more massive rope produces a smaller wave velocity.

In general, the source of a wave determines the amplitude and usually the frequency of the wave as well; but the medium in which the wave is traveling determines the velocity of the wave.*

After a while, the wave appears as shown in Fig. 9-2: What determines the distance between the peaks of the wave?

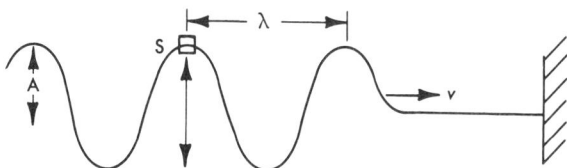

Segment S moves up and down with a certain frequency, f (in $\frac{cycles}{sec}$).
Segment S has a maximum displacement upward of A meters. This is the amplitude of the wave.
The wave moves toward the wall with a certain velocity, v (in $\frac{m}{sec}$).

The distance between peaks of the wave is called the wavelength, λ (in m).
In this case the source determines the frequency and the amplitude.
The characteristics of the rope determine the velocity.
The wavelength is determined by $\lambda f = v$.

**Fig. 9-2. Characteristics of a wave.
Relationship between the characteristics.**

The distance between peaks of the wave is called the *wavelength*, symbolized by λ, and is determined by both the frequency, f, and the velocity, v, of the wave. The relationship between these three characteristics of the wave is:

$$\text{wavelength} \times \text{frequency} = \text{velocity}$$

or,

$$\lambda f = v \tag{9-1}$$

where

λ = wavelength (m),
f = frequency (1/sec),
v = velocity (m/sec).

This formula can be remembered by simply noting the dimensions involved. Velocity is measured in m/sec, wavelength is measured in m, and

---

* Light waves are an exception to this rule; light waves do not need a medium in which to travel.

Sound

frequency is measured in 1/sec.* There is no other way of combining v, λ, and f so that the units are consistent.

Note that the amplitude of the wave is not related to the frequency, velocity, or wavelength of the wave. The source determines the frequency, the medium determines the velocity, and, finally, the above formula determines the wavelength. Sometimes the medium determines the wavelength as well as the velocity. In these cases, the frequency is determined by the above equation. For example, see Sec. 9-4, a vibrating string.

Transverse Waves and Longitudinal Waves

Suppose that I have a very long spring, and I attach one end to a very distant wall. I then continually push and pull the free end a distance $+A$ meters in and $-A$ meters out. (See Fig. 9-3.)

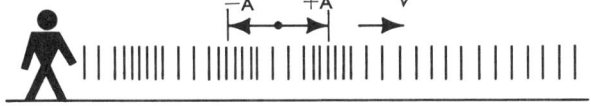

Each segment moves back and forth while the wave moves toward the right.

Fig. 9-3. A spring wave being produced by the author.

The spring is very similar to the rope of Fig. 9-2, in that each segment of spring is moving back and forth with a certain frequency and a certain amplitude. And although the average displacement of any one segment of the spring is zero, a wave, which is composed of all the segments of the spring, is moving toward the wall. The words frequency, amplitude, velocity, and wavelength have the same meaning for the wave in the rope as for the wave in the spring. Also while the frequency and the amplitude of the spring wave are determined by myself, the source of the wave, the velocity is determined by the spring itself,† and the wavelength is determined by a relationship between wavelength, frequency, and velocity. [See Eq. (9-1).]

There is, however, one important difference between the rope wave and the spring wave. As the rope wave travels toward the wall, each segment moves up and down, perpendicular to the traveling wave. But as the spring wave travels toward the wall, each segment moves back and forth, parallel to the traveling wave.

By definition, when the segments of a wave are moving perpendicular to the direction of wave travel, the wave is a *transverse wave;* and when the

---

* We talk about frequency in cycles per sec, but a cycle is not really a unit of measurement, it is an occurrence. So the unit of frequency is really 1/sec.

† The velocity of the spring wave depends on the mass and the resiliency of the spring.

segments of the wave are moving parallel to the direction of wave travel, the wave is a *longitudinal wave*. Table 9-1 presents examples of transverse and longitudinal waves.

Table 9-1 Examples of waves

Transverse waves	Longitudinal waves
Rope waves	Spring waves
Vibrating strings (e.g., violin strings)	Sound waves
Electromagnetic waves	Compression waves

9-3 SOUND

Frequency Range of Sound

Sound is a longitudinal wave because particles of air vibrate back and forth parallel to the direction of wave travel. (See Fig. 9-4.)

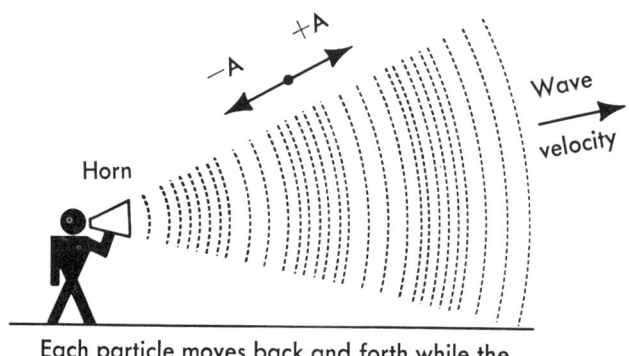

Each particle moves back and forth while the wave travels away from the source.

Fig. 9-4. Sound: a longitudinal wave.

There are three frequency ranges in which the air can vibrate. The infrasonic, the ultrasonic, and the sonic. The range of infrasonic frequencies (0 to 30 cps) is too low for human ears, the range of ultrasonic frequencies (15,000 cps to infinity) is too high for human ears, but the sonic frequencies (30 to 15,000 cps) are audible to the human ear. (See Fig. 9-5.)

The starving bat mentioned at the beginning of this chapter can emit and detect very high frequency sound waves. These waves bounce off objects, and

Sound

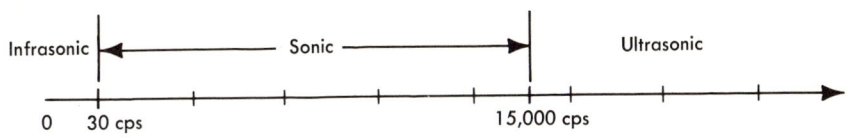

Only frequencies in the sonic range are audible to the human ear.

Fig. 9-5. **The frequency ranges of sound.**

by somehow measuring the time between emission and reflection, as well as the pattern of the reflected sound, the bat can maneuver through dense forests and satiate his appetite as well.

Velocity of Sound in Air

As mentioned in Sec. 9-2, the velocity of a wave depends on the medium through which the wave travels. In the case of sound waves, the medium is air, and so air determines the velocity of sound waves. But what are the particuliar properties of air that determine the velocity of sound waves? As it turns out, temperature determines the velocity of sound waves. As the temperature increases, the velocity increases, and vice versa. At ordinary temperatures, the velocity of sound is about 345 m/sec (1130 ft/sec).

Every word we utter travels away from our vocal cords at this enormous speed. If we are speaking in a relatively small room, the sound of each syllable bounces back and forth, from wall to wall, quite a few times even before we have completed uttering the syllable. For example, in a 20 × 20-foot room, sound can bounce from wall to wall about 55 times in one second. If a syllable lasts 0.1 second, the sound of the syllable has bounced a total of $5\frac{1}{2}$ times before the syllable ended. This is why it is easier to speak in an enclosed room than in the open air, and it also explains why voices have more "body" in an enclosed room. The constant echoing of sound from the walls and ceiling of a room enhances and enriches our voices. Eventually, of course, the energy of the sound dissipates itself into the walls, chairs, and other objects of the room. (See Fig. 9-6.)

Production of Sound

Sound can be produced in many different ways. The basic requirement is that changes in air pressure be brought about. For example, the motion of a vibrating string causes periodic changes in air pressure, and our ears can detect this periodic change. Also, the clapping of two hands produces a pulse of high air pressure, and we hear a sudden burst of sound.

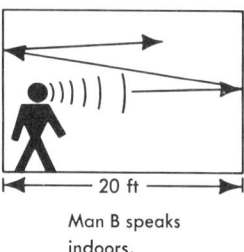

Man A speaks outdoors.

Man B speaks indoors.

Assuming that the sound wave does not lose energy as a result of its collisions with the walls of the room, it will bounce back and forth $5\frac{1}{2}$ times in 0.1 sec. These echos add "body" to a voice. Thus, man B does not have to strain his voice as much as man A.

Fig. 9-6. Speaking outdoors, speaking indoors, and echos.

Now, it can be shown that a pulse of high air pressure can be simulated by an infinite number of strings, each vibrating at different frequencies and at a different amplitude. Indeed, any kind of sound can be simulated by a sufficiently large number of strings, provided that the frequency and amplitude of each string is properly set. (See Fig. 9-7.) Let us see how a vibrating string produces sound.

9-4 A VIBRATING STRING

If a string is tied down at both ends and plucked, the string will vibrate at a certain frequency. At whatever frequency the string vibrates, the air will vibrate, and we will hear that frequency of sound. (See Fig. 9-8.) So the question is, "What determines the frequency of the vibrating string?"

Frequency of a Vibrating String

The length, tension, and mass of a vibrating string determine the frequency of sound produced. If we make the string longer or shorter, the frequency decreases or increases, respectively. If we make the string looser or tauter, the frequency decreases or increases, respectively. Finally, if we make the string lighter or heavier, the frequency increases or decreases, respectively.

The exact sequence of events that determines the frequency is as follows: Once the string is plucked,

1. the length of the string determines the wavelength, and
2. the tension and mass of the string determine the velocity of the wave on the string, and finally,
3. the frequency of the string is determined by $\lambda f = v$.

Sound

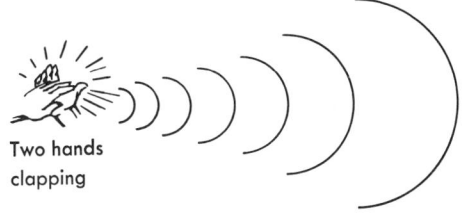

Two hands clapping

Pulse of high air pressure producing sound

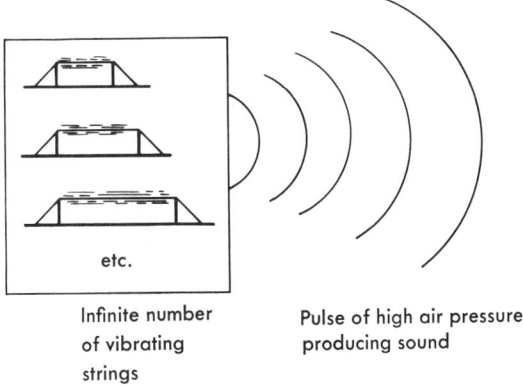

Infinite number of vibrating strings

etc.

Pulse of high air pressure producing sound

Any sound can be produced by superimposing a sufficient number of vibrating strings. Each string vibrates with a particular frequency and with a particular amplitude.

Fig. 9-7. **The equivalence of vibrating strings and sound.**

Fundamental Frequency, Harmonic Frequencies

There is a slight complication to this process; although the tension and mass of the string determine just one wave velocity, the length of the string doesn't determine just one wavelength. The length of the string determines a *class of wavelengths* (or a *wavelength pattern*). And since each wavelength in the wavelength pattern has associated with it a certain frequency, the string vibrates in a certain frequency pattern.

Let's look at a particular example. We will determine the wavelength pattern of a string and, from the wavelength pattern, we will determine the frequency pattern.

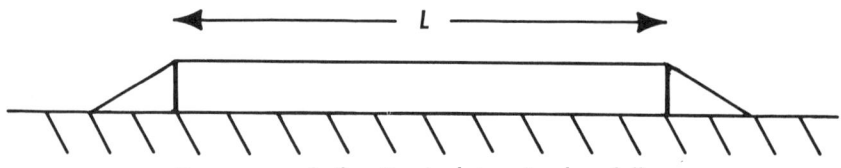

Frequency of vibration is determined as follows:
(1) length L determines λ
(2) the string itself determines v, and
(3) the frequency is such that $\lambda f = v$.

Now if the string is vibrating at a frequency f, the air will vibrate at a frequency f. Therefore, we hear a sound of a frequency f.

Fig. 9-8. A vibrating string.

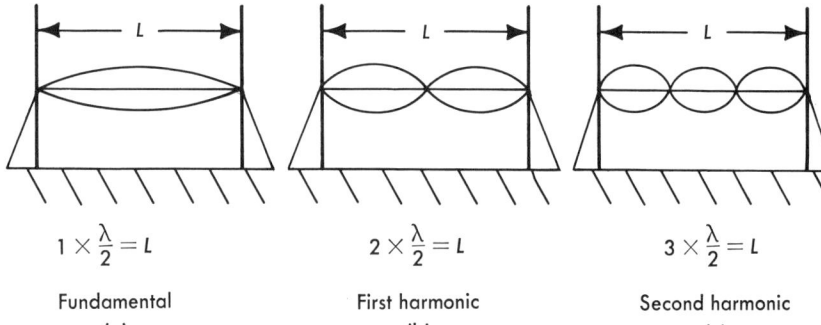

$1 \times \dfrac{\lambda}{2} = L$ 　　　　$2 \times \dfrac{\lambda}{2} = L$ 　　　　$3 \times \dfrac{\lambda}{2} = L$

Fundamental 　　　　 First harmonic 　　　　 Second harmonic
(a) 　　　　　　　　　　(b) 　　　　　　　　　　(c)

The wavelengths, λ, coupled with the velocities, v, produce various frequencies.

Fig. 9-9. Fundamental frequency, harmonic frequencies.

If we have a string of length L [see Fig. 9-9(a)], one of the possible wavelengths is determined by the fact that

$$1 \times \frac{\lambda}{2} = L$$

or

$$\lambda = 2L$$

Another possible wavelength [see Fig. 9-9(b)] is

$$2 \times \frac{\lambda}{2} = L$$

or

$$\lambda = \frac{2L}{2}$$

Still another possible wavelength [see Fig. 9-9(c)] is

$$3 \times \frac{\lambda}{2} = L$$

Sound

or

$$\lambda = \frac{2L}{3}$$

In general, we see that the class of wavelengths allowed by a string of length L is

$$\lambda = \frac{2L}{1}, \text{ or } \frac{2L}{2}, \text{ or } \frac{2L}{3}, \text{ or } \frac{2L}{4}, \ldots$$

This class of wavelengths constitutes the wavelength pattern of the string.

Now, for each of the wavelengths in the wavelength pattern, there is a corresponding frequency. The corresponding frequency is determined by Eq. (9-1), $\lambda f = v$. For example, if $v = 400$ m/sec and $L = 1.0$ m,

Allowed wavelength	Allowed frequencies ($\lambda f = v$)
$\lambda_1 = \dfrac{2L}{1} = 2.00$ m	$2.00 \times f_1 = 400$ or $f_1 = 200$ cps
$\lambda_2 = \dfrac{2L}{2} = 1.00$ m	$1.00 \times f_2 = 400$ or $f_2 = 400$ cps
$\lambda_3 = \dfrac{2L}{3} = 0.67$ m	$0.67 \times f_3 = 400$ or $f_3 = 600$ cps
$\lambda_4 = \dfrac{2L}{4} = 0.50$ m	$0.50 \times f_4 = 400$ or $f_4 = 800$ cps
$\lambda_5 = \dfrac{2L}{5} = 0.40$ m, etc.	$0.40 \times f_5 = 400$ or $f_5 = 1000$ cps, etc.

Notice that the frequency pattern can be written as follows:

$$f = \frac{v}{\lambda_1}, \text{ or } \frac{v}{\lambda_2}, \text{ or } \frac{v}{\lambda_3}, \text{ or } \ldots$$

$$= \frac{v}{\frac{2L}{1}}, \text{ or } \frac{v}{\frac{2L}{2}}, \text{ or } \frac{v}{\frac{2L}{3}}, \text{ or } \ldots$$

$$= v \times \frac{1}{2L}, \text{ or } v \times \frac{2}{2L}, \text{ or } v \times \frac{3}{2L}, \text{ or } \ldots$$

$$= \frac{v}{2L} \times 1, \text{ or } \frac{v}{2L} \times 2, \text{ or } \frac{v}{2L} \times 3, \text{ or } \ldots$$

$$= 200 \times 1, \text{ or } 200 \times 2, \text{ or } 200 \times 3, \text{ or } \ldots$$

$$f = 200 \text{ cps, or } 400 \text{ cps, or } 600 \text{ cps, or } \ldots$$

144   *Mechanics, Heat, and Sound*

The first allowed frequency (200 cps), is called the *fundamental frequency,* and the others (400, 600, . . .) are called *harmonics,* or overtones, of the fundamental frequency.

In general, when a particular string is plucked, the fundamental frequency as well as all of its harmonics are produced simultaneously, causing a complex frequency pattern. The air then vibrates in exactly the same complex frequency pattern, and we hear sound.

Importance of Harmonic Frequencies

Even though two different strings have the same fundamental frequency, if they are plucked differently or arranged differently, a different set of harmonics may be produced by each string. The ear is a very complex device and is able to detect very slight differences in frequencies and frequency patterns. (See Fig. 9-10.) For example, the musical note A above middle C is defined to be 440 cps, but A above middle C on the piano produces a note different from A above middle C on a violin. The difference in sensation is

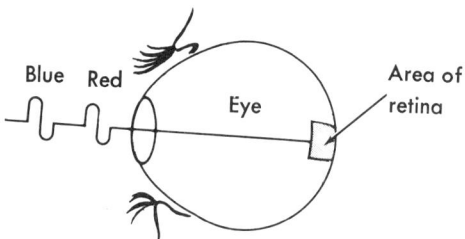

Red light and blue light strike the same area of the retina. We see violet light (a mixture of red and blue).

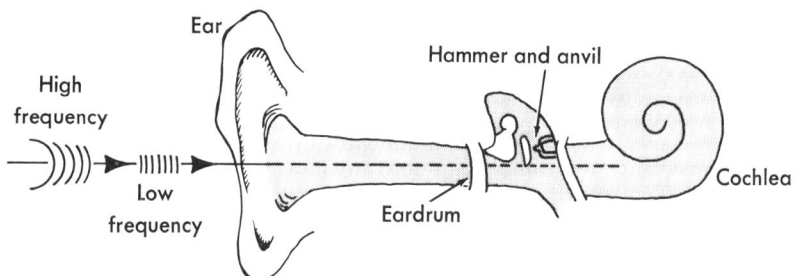

A low-frequency sound wave and a high-frequency sound wave strike the eardrum. The cochlea can separate the two frequencies and we can sense both the low and the high frequencies separately.

Fig. 9-10. Schematic diagram of the eye sensing light waves. Schematic diagram of the ear sensing sound waves.

Sound

caused by the difference in the number and intensity of harmonics produced by the piano as compared to those produced by the violin.

9-5 THE EYE AND THE EAR

Electromagnetic Waves and the Eye; Sound Waves and the Ear

When you study electricity and magnetism, you will learn that light is an electromagnetic wave, and that the different colors of light correspond to different frequencies of electromagnetic waves. For example, red light is an electromagnetic wave of frequency 3.5×10^{14} cps, and blue light is an electromagnetic wave of frequency 7.0×10^{14} cps. Just as the eye is sensitive to light waves, the ear is sensitive to sound waves. And just as the eye is sensitive to only a certain range of light waves, the ear is sensitive to only a certain range of sound waves. But there is a basic difference between the eye and the ear! (See Fig. 9-11.)

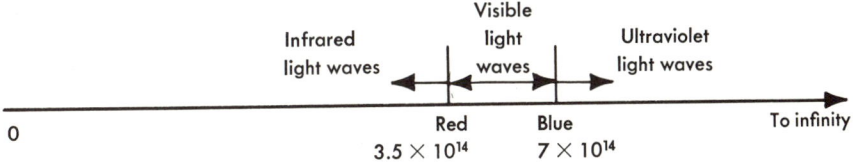

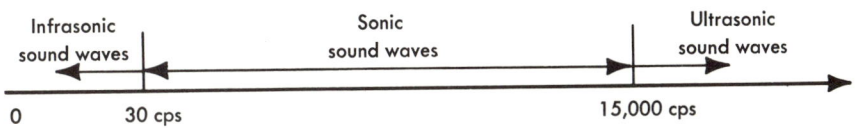

Fig. 9-11. Frequency spectrum of light waves.
Frequency spectrum of sound waves.

What happens if two different frequencies of light simultaneously strike the same area of the retina? If one frequency is that of red light, and the other is that of blue light, we will sense a magenta sort of color. Our eyes can only sense a superposition of the two colors; they cannot distinguish the red from the blue. But what happens if two different frequencies of sound strike the ear drum?

Within the ear is a very sensitive device, the cochlea, which can separate the two different frequencies; consequently, we hear each frequency separately. That is to say, our ears can break up a complex wave pattern into its various frequency components. (See Fig. 9-10.) In this way, given an assortment of sound, we can pick out the frequency pattern we choose to hear; i.e., in a crowded, noisy room we are often able to listen to the person of our choice, because we can "tune in" to his frequency pattern and "drop out" other frequency patterns.

9-6 REVIEW

Waves are characterized by their frequency, wavelength, velocity, and amplitude. There is a relationship between the frequency, wavelength, and velocity; namely, $\lambda f = v$.

Sound is a longitudinal wave, and its frequency can range from 0 cps to infinity. Our ears are sensitive to only a finite region of this spectrum (about 30 to 15,000 cps).

Bats use sound waves as we use our eyes. Using pulses of high-frequency sound, they are able to "see" trees, branches, mosquitoes, etc. The frequency range used by bats is too high for human ears to hear.

A vibrating string is an excellent example of a sound-producing device. Properly plucked, a string can generate a rich combination of fundamental and harmonic frequencies. The fundamental frequency is determined by the length, tension, and mass of the string. The harmonics are, of course, multiples of the fundamental.

Although the eye is sensitive to light waves and the ear is sensitive to sound waves, their sensitivities are basically different. The eye responds to a superposition of *all* of the light waves impinging on a particular spot of the retina, while the ear can decompose the sound wave into its various frequency components. In this way, we can concentrate on the treble frequencies if we enjoy them more than the bass frequencies, and vice versa.

PROBLEMS

1. If the period of a wave is 0.01 sec, what is its frequency?
2. Discuss the difference between wavelength and amplitude. (Use pictures and words.)
3. The velocity of a wave depends on which of these factors: period, frequency, amplitude, wavelength, or the medium in which the wave travels?
4. What is the difference between a transverse wave and a longitudinal wave? Give an example of each.
5. A wire is stretched between two posts. The tension and mass of the wire is such that the velocity of a wave on the wire is 500 m/sec. If the two posts are 2 m apart, what would the fundamental frequency of the wire be? What are the frequencies of the first 3 harmonics?
6. In Problem 5, if we wanted to lower the fundamental frequency slightly, should we increase or decrease the tension in the wire?
7. In Problem 5, how can we double the fundamental frequency, using the same wire with the same tension?

Appendix I

A FALLING BODY

In Fig. 1-9, a marble is dropped from a height h, and the problem is to relate the height h to the initial velocity, the final velocity, the acceleration, and the time of travel of the marble. Since the marble is undergoing constant acceleration; i.e., its velocity is changing at a constant rate, we cannot use the equation $d = vt$. But perhaps we can find an average velocity for this marble; then we can say

$$h = v_{\text{average}} t \qquad \text{(A1-1)}$$

where h is the distance traversed, v_{average} is the average velocity, and t is the time of travel.

What is the average velocity? The average velocity is, as you might expect, the final velocity plus the initial velocity divided by two; i.e.,

$$v_{\text{average}} = \frac{v_{\text{final}} + v_{\text{initial}}}{2} \qquad \text{(A1-2)}$$

where v_{final} is the final velocity, and v_{initial} is the initial velocity.

Since the only things we know about this marble are its initial velocity (v_{initial}) and its acceleration (a), we must try to write the final velocity (v_{final}) in terms of v_{initial} and a. Then we can determine the average velocity in terms of v_{initial} and a. Using the definition of acceleration [Eq. (1-2)], we get

$$a = \frac{v_{\text{final}} - v_{\text{initial}}}{t} \qquad \text{(1-2)}$$

multiplying by t,

$$at = v_{\text{final}} - v_{\text{initial}}$$

adding v_{initial},

$$v_{\text{initial}} + at = v_{\text{final}}$$

Using this expression in Eq. (A1-2), we obtain

$$v_{\text{average}} = \frac{v_{\text{initial}} + at + v_{\text{initial}}}{2}$$

$$= \frac{2v_{\text{initial}} + at}{2} \quad \text{or} \quad \frac{2v_{\text{initial}}}{2} + \frac{at}{2}$$

$$v_{\text{average}} = v_{\text{initial}} + \frac{at}{2} \qquad \text{(A1-3)}$$

At last we are in a position to find the distance traveled by this marble over a period of time, t. Using Eq. (A1-1),

$$h = v_{\text{average}} \times t$$
$$= \left(v_{\text{initial}} + \frac{at}{2}\right) \times t$$
$$h = v_{\text{initial}}t + \frac{at^2}{2} \tag{A1-4}$$

Appendix II

VECTOR ADDITION AND SUBTRACTION

A *scalar* is a physical quantity that requires only a magnitude for its description; e.g., temperature. However, a *vector* is a physical quantity that requires both a magnitude and a direction for its description; e.g., distance traveled. The addition and subtraction of vectors is more complicated than the addition and subtraction of scalars. Let us consider a particular case of adding and subtracting vectors.

Vector A: 5 ∡ 45°

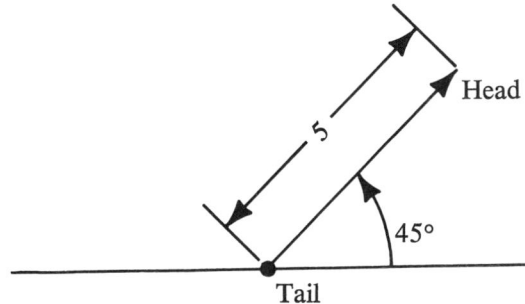

Vector B: 8 ∡ 30°

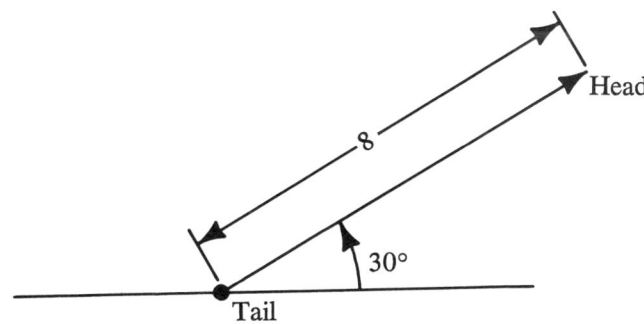

Arithmetic Method

	x-direction	y-direction
$A = 5 \angle 45°$	3.54	3.54
$B = 8 \angle 30°$	6.93	4.00

Addition: $A + B$		Subtraction: $A - B$	
x-direction	y-direction	x-direction	y-direction
3.54	3.54	3.54	3.54
6.93	4.00	−6.93	−4.00
Sum 10.47	7.54	Sum −3.39	−0.46

We will use the arithmetic method and the x- and y-portions of a vector will always be given. For completeness, the graphic method is also given.

Graphic Method (Choose a scale; e.g., let $\tfrac{1}{4}'' = 1''$.)

Addition: $A + B$	Subtraction: $A - B = A + (-B)$
	[Definition: To obtain the negative of a vector, add 180°. For example, $-(8 \angle 30°) = 8 \angle 210°$.]

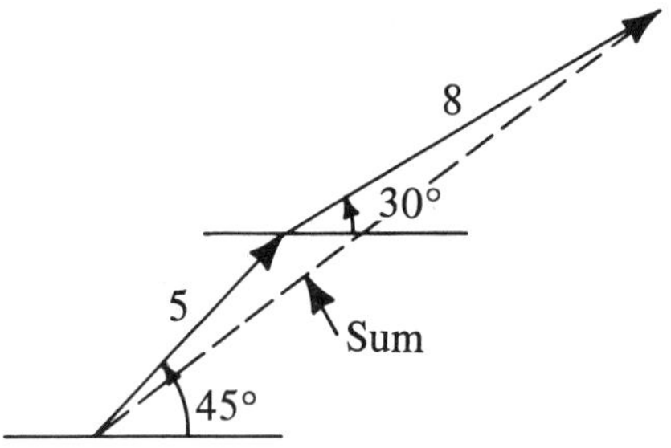

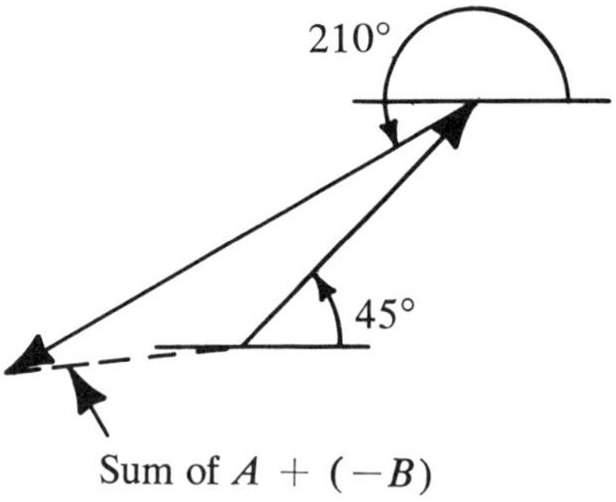

Sum

Sum of $A + (-B)$

Using a ruler and protractor, you will find that this sum is $13 \angle 37°$.	Using a ruler and protractor, you will find that this sum is $3.5 \angle 188°$.

Add vectors by connecting them head-to-tail. The sum is that vector which goes from the very beginning of your diagram to the very end of your diagram.

Appendix III

PROBLEMS WITH THE CONCEPT OF A STRAIGHT LINE

A straight line is the shortest distance between two points. Let us imagine people living in two different two-dimensional worlds.

I	II
One two-dimensional world is a flat plane, like that of a tabletop, extending to infinity in all directions. A straight line to people living on this "tabletop" world appears to us* to be a normal straight line.	The second two-dimensional world is a circular plane; i.e., the surface of a large sphere. A straight line to people living on the surface of this "spherical" world appears to us as an arc of a great circle; i.e., a straight line to them is not a normal straight line to us.

Notice two things:

1. To the people who live in either of these two-dimensional worlds, there is no question as to the meaning of a straight line.
2. To clearly and easily see the shape of the straight line, we had to view it from a higher than two-dimensional world; the line had to be viewed from a three-dimensional world.

Indeed, it is the shape or the *geometry* of the two-dimensional world in which these people live that determines the form of the straight line as viewed by us, who live in a three-dimensional world. Is it possible for these people to discover the geometry of their two-dimensional world by themselves? Yes, if they perform some very accurate measurements.

For example, if the geometry of the two-dimensional world is a flat plane, then the sum of the angles of a triangle is 180°; but if the geometry of the two-dimensional world is spherical in shape, the sum of the angles of a triangle is greater than 180°. (See Fig. A3-1.) If these people would measure very accurately the sum of the angles in a triangle in their two-dimensional world, they would be able to deduce the geometry of their world. In a similar way, the geometry of our own three-dimensional world can be discovered by performing very accurate measurements.

Newton, as well as everyone else, felt that we live in a three-dimensional world and that time flowed along independent of and separate from this three-dimensional world.

---

* We live in a three-dimensional world, and are viewing these people in their two-dimensional world. Note that since time exists in the two-dimensional world as well as in our three-dimensional world, we will disregard time in both worlds at the moment.

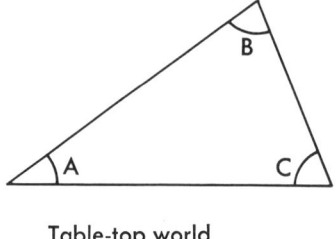
Table-top world
Angles A + B + C = 180°

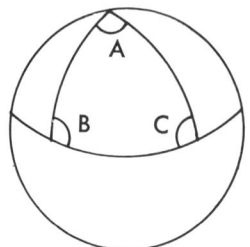
Spherical world
Angles A + B + C > 180°
(Since angles B and C
are already equal to 90°)

Fig. A3-1. **Two different 2-dimensional worlds as viewed from a 3-dimensional world.**

Also, Newton, as well as everyone else, felt that a straight line looks exactly like what we normally call a straight line. In 1905, however, Einstein's special theory of relativity showed that no one had ever really understood time. (It is not certain that anyone understands it today, either.) Time is closely tied to space, and as soon as we include time we must add another dimension to our worlds. Thus, our hypothetical two-dimensional worlds are really three-dimensional space-time worlds, and our own three-dimensional world is really a four-dimensional space-time world. So we are actually moving in a four-dimensional space-time world. (Even when we are standing still, we are moving in time.)

The key question is, "What does a straight line look like in a four-dimensional world?" Like the straight lines in our tabletop world and in our spherical world, it depends on the geometry of our four-dimensional space-time world. According to Einstein's general theory of relativity, that geometry depends on the masses in the region. Unfortunately, there is not enough space in this book to discuss Einstein's theories, so I must refer the reader to another text.*

---

* See Isaac Maleh, *Modern Physics,* Charles E. Merrill Publishing Co., Columbus, Ohio, 1966.

Index

Index

Absolute inertial frame of reference, 25
Absolute zero, 119-121
   determination of, 120
Acceleration, 11-13
   centripetal, 21
   definition, 11
   due to gravity, 46
Ammonia gas, 108
Amplitude, wave, 135
Angular momentum, 89-90
   conservation, 89
   definition, 89
Atomic theory and J. Dalton, 106
Atoms, 104-112
Avogadro, A., 108
   atomic hypotheses of, 108-109
Avogadro's number, 112

Boltzmann's constant, 130
Boyle, Robert, 126
Boyle's law, 126

Calorie:
   conversion factor, 92
   definition, 90
Cavendish experiment, 43
Cavendish, Henry, 43
Celestial sphere, 6
Centripetal acceleration, 21
Centigrade temperature scale, 119
Charles, J. A. C., 126
Charles's law, 126

Circular motion, 19-21
   direction of $F, v,$ and a, 32
   summary, 55
Compound, 103-106
   definition, 104
   empirical definition, 105
Conservation of momentum, 83-88
   example, 84
   statement, 83
Conservation of angular momentum, 89-90
   example, 90
   statement, 89
Conservation of energy, 90-99
   example, 94
   statement, 91
Copernican system of universe, 7, 76

Dalton, John, 106
Deferent circle, 6
Descartes, 10, 34
Dyne, definition, 32

Earthly bodies, 1-6
Einstein, A., 72
Ellipse, definition, 53
Electric charge, conservation of, 99
Elements, 101-108
   basic, 1
   definition, 104
   empirical definition, 105
   nature, 104
   natural position, 4

Energy, 90-99
 conservation of, 91
 forms of, 91
 heat, 97
 kinetic, 91, 93-98
 potential, 91, 92-93, 95-98
 units, 92
Energy flow and temperature, 121
English system of units, 32
Epicyclic circles, 6
Equivalence, principle of, 50-51
Ether, 25
 and the gravitational force, 47

Fahrenheit temperature scale, 118
Fahrenheit, G. D. and mercury thermometers, 118
Force:
 centrifugal, 69
 centripetal, 69
 CGS unit of, 32
 MKS unit of, 31
 moving, 76
Forces, saturation of, 74
Four-dimensional space-time, 77, 152
Frame of reference:
 absolute inertial, 25
 inertial, 25
Free fall, 66
Frequency of a wave, 135
Fundamental frequency, 141

g, 46
G, measurement, 44
G, universal gravitational constant, 43
Galilean thermometer, 116
Galileo, 9-10, 34
 thermometry and, 116
Gas, ideal:
 definition, 126
Gases, kinetic theory of, 124-132
Gaseous state, 112
 description, 113

General relativity:
 and Mercury, 72
 brief summary, 77
Geometry and space-time worlds, 152
Gravitation:
 brief history, 76-78
 law of, 39-49
Gravitational field, 47-49
Gravitational force, 39-49
 definition, 40-42
 mechanism, 80
 non-saturation, 74
 range, 40
 saturation, 73-74
 source, 47-49
 speed, 73
Gravitational mass, 50-51
Gravitons and gravitational forces, 47

Harmonic frequencies, 141-145
 importance of, 144
Heat and temperature, 121
Heavenly bodies, 1-6
Homogeneous, 103
Hydrogen, composition of, 104
Hydrogen chloride gas, 108
Huygens, 10
 and thermometry, 118
 and circular motion, 20

Ideal gas, definition, 126
Inertia, 24
Inertial mass, 37, 50-51
Inertial frames, 25-28
 examples, 27

Joule, unit of energy, 92

Kelvin temperature scale, 119-120
Kepler, Johannes, 53
Kepler's system of the universe, 7
Kepler's laws, 53-54
Kinetic energy, 91, 93-98
 and temperature, 130

Index

Kinetic theory of gases, 124-132
    hypotheses, 127

Lavoisier, A., 105
Liquid state, 112-113
    description, 113
Longitudinal waves, 137-138
    definition, 138
    examples, 138

Mass, 24
    gravitational, 50-51
    inertial, 37, 50-51
    inertial and gravitational, a comparison, 50-51
Mathematical Principles of Natural Philosophy, 23
Matter, 101-122
    atomistic nature of, 106
    composition, 101
    states, 112
        factors determining, 114
Mercury and general relativity, 72
Mercury thermometer, 118
Mesons and nuclear forces, 49
Milky Way galaxy, 26, 78
Mixture, 103-105
    definition, 104
    empirical definition, 105
MKS system of units, 31
Mole, definition, 112
Molecular weight, 112
Molecules, 104-112
Motion, 1-37
    Aristotelian theory, 1-7
        downfall of, 7
    brief history, 76-78
    circular, 19-21
        summary of, 55
    earthly bodies, 1-6
    first law, 24-29
        statement of, 24
    heavenly bodies, 1-7
    importance of initial conditions, 35-36

Motion—*cont.*
    natural,
        earthly bodies, 5
        heavenly bodies, 4
    Newton's laws of, 23-35
    second law, 29-33
        statement of, 29
    third law, 33-34
        statement of, 33
    of three bodies, 61
    uniform, 35
    unnatural, 5
Momentum, 83-88
    conservation of, 83
    definition, 83
    units of, 85
Moving force, 5

Neap tides, 69
Neptune, discovery of, 72
Newton, Sir Isaac, 10, 23
Newton:
    unit of force, 31
        conversion to lb, 32
Newton's law of gravitation, 39-49
Newton's laws and Kepler's laws, a comparison, 59-60
Newton's laws of motion, 23-38
Newton's laws, predictions of, 58
Non-inertial frames, 27-28
    examples, 27-28
North Star, 2, 6

Orbits, class of, 58
Overtones and harmonics, 144

Parity, conservation of, 99
Perfect gas law, 124-127
Period of a planet, definition, 54
Period of a wave, definition, 135
Perturbation theory, 71
Perturbations, 70-72
Photons and electromagnetic forces, 49

Pluto, discovery of, 72
Poker, conservation laws and, 81
Potential energy, 91-97
Pounds, lb, conversion to newtons, 32
Pressure, relation to V and T of gas, 125-127
Principia, 23
Principle of equivalence, 9, 51
  consequence of, 58
Ptolemaic system of universe, 1, 76

Quantum theory, 60

Relativity theory, 60

Saturation of forces, 74
Scalars, 15-16
  examples, 16
Second law of motion, 29-33
  example, 30
  statement, 29
Slug, 32
Solid state, 112-113
  description, 113
Sound, 133-146
  frequency range, 138
  fundamental frequency, 141
  harmonic frequency, 141
  infrasonic frequencies, 138
  production, 139-140
  sonic frequencies, 138
  ultrasonic frequencies, 138
  velocity in air, 139
  vibrating strings, 140
Space-time, 77
Space-time worlds, geometry of, 152
Speed, definition, 19
Spherical world, 151
Spring tides, 69
Standard temperature and pressure, 112
STP, 112
Straight lines, 25, 151

States of matter, 112-115
  and temperature, 115
Substance, definition, 103
System of units:
  CGS, 32
  English, 32
  MKS, 31

Tabletop world, 151
Temperature, 115-123
  and energy flow, 121
  and heat, 121
  measurement, 115
  relation to P and V of gas, 125
  thermometry, 115-121
Temperature scales, 118-121
  absolute, 120
  centigrade, 119
  Fahrenheit, 118
  Kelvin, 120
Thermometer:
  Galilean, 116-117
  mercury, 117
  gaseous, 119
Thermometry, 115-121
Thomson, W. and Baron Kelvin, 120
Tides, 68-70
  spring tides, 69
  neap tides, 69
Transverse waves, 137-138
  definition, 138
  examples, 138

Uniform motion, 35
Universal gravitational constant, G, 43
Universe:
  Copernican system, 7, 76
  Keplerian system, 7
  Ptolemaic system, 1, 76

Vectors, 14-17
  examples, 15-16
  table of, 16

Index

Velocity, 10
  as a vector, 20
Vibrating string, 140-144
  fundamental frequency, 141
  frequency pattern, 143
  harmonic frequency, 141
  wavelength pattern, 141, 143
Volume, relation to P and T of gas, 125

Water:
  composition, 104
  volume experiments, 108
Waves, 133-146
  amplitude, 135

Waves—*cont.*
  characteristics, 133-136
  examples, 138
  frequency, 135
  longitudinal, 137-138
  period, 135
  transverse, 137-138
  velocity, 136-137
  wavelength, 136
Weight:
  definition, 45
  and weightlessness, 63
Work, definition, 91

Zero, absolute, 119-121